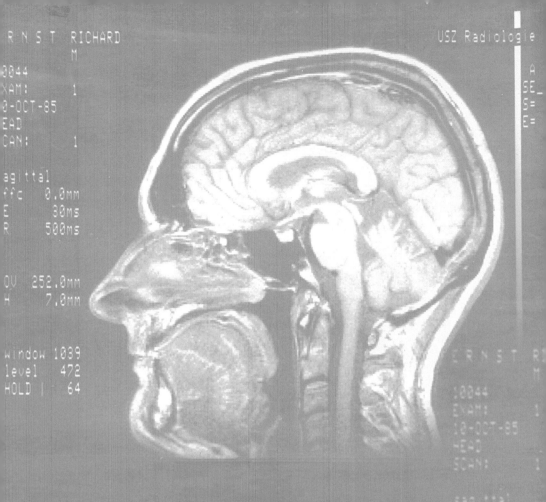

Searching and Researching

Searching and Researching

An Autobiography of a Nobel Laureate

Richard R. Ernst *with* Matthias Meili

Translated by Mark Pearce

JENNY STANFORD
PUBLISHING

Published by

Jenny Stanford Publishing Pte. Ltd.
Level 34, Centennial Tower
3 Temasek Avenue
Singapore 039190

Email: editorial@jennystanford.com
Web: www.jennystanford.com

British Library Cataloguing-in-Publication Data
A catalogue record for this book is available from the British Library.

Searching and Researching: An Autobiography of a Nobel Laureate

Translated by Mark Pearce

ISBN 978-981-4877-92-3 (Hardcover)
ISBN 978-1-003-20030-7 (eBook)

Stockholm on Pan Am 31

It is 16 October 1991 and I am sitting on a flight to New York from Moscow, where I had delivered a lecture. I am heading to New York to receive the Louisa Gross Horwitz Prize. I feel great pride in being awarded such an illustrious, international prize, even if few outside the research community have heard of it. We are three hours into the flight, somewhere between Scotland and Ireland, and I gaze out of the window at the plane's engines humming, set against the endless blue sky. I am sitting in business class, and as is always the case on a flight, am preparing my next lecture. Over the top of my notes, I suddenly see the pilot enter the passenger cabin and I start to think: Is this an emergency or just a routine check? It is neither. The captain approaches me and leans down to my seat. "Mr. Ernst?" he asks. I nod. "Could you please come with me into the cockpit. We have a call for you. From Stockholm." Stockholm? This can only mean one thing: the Nobel Prize!

My sister recently told me that when I was young, she always used to tease me by saying I would at some point win this accolade like no other. In other words, I was seen as a bit of a "nerd" even then, a reputation that was not necessarily the envy of my peers. But for scientists, there is no higher honor than a Nobel Prize. It is impossible to describe the sense of satisfaction and pride that it brings. Aside from the honor of receiving one, there is also the matter of the sizable cash prize – something that is of course nice to receive, yet still somehow unexpected. And when it is your turn and you take the call, you seize the prize with both hands. It goes without saying I am overjoyed, but at the same time, I feel a pang of conscience: Have I really earned it given science is ultimately teamwork? Who else has won one, who hasn't? What will other people think? These are just some of the thoughts going round my brain as I make my way to the cockpit.

Picking up the aircraft radio quickly brings me back down to earth. On the other end of the call is the Secretary General of the Royal Swedish Academy of Sciences, who gives me the good news and congratulates me. A call is then quickly placed to Zurich, where a spontaneous press conference is being set up in my honor. Although there is some interference on the line, I can make out the voice of Jakob Nüesch, the President of the Swiss Federal Institute of Technology in Zurich (ETH Zurich), who passes on his congratulations. It is then the turn of the assembled journalists to ask me their questions: How do I feel? What will I do with the prize money? While Swiss journalists have priority, I hear someone speaking broken Swiss-German with an Italian accent. I assume that it must be a journalist from the Italian-speaking south of the country. "Mr. Ernst. This is Flavio Cotti on the line. *Gratulazione.* You are a credit to our country!" It turns out I am mistaken: I am now talking to the President of Switzerland. It is at this moment that it begins to sink in and a feeling of joy begins to overcome me. I think of my mother Irma, my wife Magdalena, my children Anna, Katharina, and Hans Martin, all sat at home in Winterthur. It has been many years since they accompanied me on my travels for my scientific career. I told Magdalena a long time ago, on our wedding day even, that I would never have a great deal of time for family – and she accepted this graciously. And now, at the moment of my greatest success, this is something I regret. It is bizarre that while people need their loved ones when they are suffering, they need to be with them even more during those moments of pure elation. As the Swedish proverb goes, a joy shared is a joy doubled. I am unable to contact my family as the radio connection is once again interrupted.

I make my way back to my seat, somewhat bemused by what has gone on. I am met there by the cabin crew, who ask for a souvenir photo. I feel like a cyclist standing on the top of a podium, surrounded by pretty air stewardesses. Can this really be true? In my head, I replay the call from Stockholm, trying hard to work out if there has been a mistake. I dissect each and every word of the Committee's motivation for awarding me the Nobel Prize in Chemistry: "for my contributions to the development of the methodology of high resolution nuclear magnetic resonance (NMR) spectroscopy". Originally developed as an analytical method to be used in the field of chemistry, the importance of NMR spectroscopy now extends far beyond this. Nowadays you will find an MRI scanner in the basement of any hospital, with thousands of patients and test subjects being scanned each and every day around the world. The MRI images produced help detect dangerous illnesses, identify brain tumors at an early stage, and show where blood vessels are clogged – in short, they save lives.

This short explanation does a disservice to all the brilliant scientists who also contributed to the development of this highly effective method, without whom I would not have been in a position to carry out my work. Many of the Nobel Prize recipients who went before me won the prize for developing theoretical principles, which I then built on. But a great many researchers who came after me and made key contributions towards transferring methods from chemistry labs into hospitals have gone unrecognized. Or the colleagues I have worked alongside, who have not received (or are still yet to receive) any commendation for their work. Wes Anderson, for example: My friend and previous boss at Varian Associates in Palo Alto, California. In the 1960s, Wes and I spent many days and nights discussing and tinkering with various processes until we found a way to make nuclear magnetic resonance a functional method. And my colleagues at ETH Zurich: "What happened to Kurt Wüthrich?" I was later quoted as saying in the newspapers. Did the Committee in Stockholm simply overlook him? The man who applied the method to research and then foster an understanding of macromolecules? When I soon afterwards realized that I was the sole recipient of the prize, I felt a sense of embarrassment.

Richard Ernst celebrates winning the Nobel Prize with the cabin crew shortly after receiving the famous phone call from Stockholm in mid-flight from Moscow to New York.

I had barely landed in New York and I saw that another press conference had been set up in the arrivals hall of JFK Airport – was this also for me? Or maybe because I was coincidentally on the last Pan Am flight to ever take off? On this day of all days, it so happened that this long-established US airline had gone bankrupt. Later on in the hotel, I meet Kurt Wüthrich, with whom I had jointly received the Louisa Gross Horwitz Prize as mentioned above. Together we had done it: We had managed to turn ETH Zurich into an international mecca for research into nuclear magnetic resonance. Him, the ambitious sports teacher, the high flyer. And me, the silent achiever, who frequently used to doubt himself. But our meeting in New York is somewhat awkward, given our working relationship has been, to put it mildly, troubled for a number of years now. It is a relief to be able to leave New York again. Happily, eleven years later Kurt would also go on to receive the Nobel Prize in Chemistry for his development of nuclear magnetic resonance spectroscopy in the area of biomolecules. While this represented another success for ETH, more importantly it helped thaw the frosty relationship that we had after he was overlooked in 1991.

Top scientists can be strange people, and I include myself in this. To achieve success, you have to exercise great discipline over your own needs; you have to do everything to achieve your goal in the name of science. There is no room here for emotions, feelings, indeed the "soul" of the scientist. The single overarching goal is to be as "objective" as possible in representing the laws of nature. That is why scientists frequently neglect to exercise their personal freedoms in many respects. Nevertheless, they are still people, working away in a lab, with all their emotional ups and down, their irrational tendencies, which at first glance would only seem to have a disruptive effect on objective science. However, I am convinced that such a multi-faceted personality is absolutely essential to progress. As they say, a person who stands on only one leg can rarely move forwards quickly. At first it was classical music, then the Tibetan art that became extremely important in my life. I immersed myself in the Buddhist culture, building up a collection of Tibetan scroll paintings, the so-called thangkas. They provided a balance in my life and became a real passion, something that helped me overcome any number of crises. Overall my life has been an emotional rollercoaster, full of highs and lows. I never had the feeling that I was a lucky person; fate never seems to favor me, but despite all the hurdles I have faced, I have done my own thing. But never did I want to be known just as "the scientist" or the "spectroscopist".

Contents

Childhood and Youth, 1933–1952

Difficult beginnings

When I was young, my thoughts were dominated by a search for the truth and I was fascinated by the laws of nature. In those days, and indeed in days to follow, I was a lonely person, often keeping my distance from others – something that I find hard to explain. Especially if you consider my background and the almost perfect circumstances in which my life began.

I was born on 14 August 1933 into a traditional family in Winterthur. At the time this was a cause for celebration – I was the firstborn, the son and heir. My father was an architect and professor at what was then known as the Technikum Winterthur, the School of Engineering, and my mother was a housewife. We lived in a late-19th-century brick stone villa, with a large wooded garden, right next to the railway line heading to eastern Switzerland. The Ernst family tree can trace its roots back to the 16th century, with all its various minutiae documented in an impressive family chronicle. Being born into such a family is as much a responsibility as it is a privilege.

In those days, Winterthur had a long-standing reputation as a town of business and industry, whose wealth was primarily built on the cotton trade – a pattern familiar throughout industrialized Europe. On the back of the demand for machinery to be used in the textiles industry, the Sulzer and Rieter engineering works grew to become national icons of the industrial age. For many decades, the Swiss Locomotive and Machine Works produced its distinctive dark-green trains featuring the Swiss cross, which journeyed throughout the country and helped forge Switzerland's travel links. The Volkart and Reinhart families built up international trading empires, bringing prosperity to the town, which was quickly becoming a hub for art and culture, as well. Banks and insurance companies were established, bridging the gap between townspeople keen to invest their money and factories on their way up in the world. My ancestors were involved in a number of such companies. My grandfather Walter Ernst was in the iron business, an industry that underpinned the dawning industrial age. His brother Rudolf Ernst-Reinhart was an engineer and partner in the Sulzer engineering works and sat on the Board of Directors of the "Bank in Winterthur", which would go on to become what is now the major bank UBS.

Mother Irma Ernst-Brunner and father Robert Ernst with one-year-old Richard in the garden of their house in Winterthur, around 1934.

So while I was born into a family that had a certain status, my beginnings were anything other than straightforward. Until the age of three, I refused to speak an intelligible word. The only person who could understand me was my sister Verena, who was born a year after me. Throughout my early years, Verena was my loyal companion through thick and thin. She was the only one who could understand the secret language that I had developed for myself. It is extraordinary to think that I can still remember some words now. "Ma peng gi-ga-gi" meant "soldiers", for example, while "gi-ga-gi" meant "lots". Barely out of nappies herself, Verena understood everything I said, translating this for my parents, my grandparents and anyone else who would try to talk to this unique little boy. I rewarded her with all manner of strange grimaces and giggles, sometimes making her laugh so hard she nearly wet herself! This was not easy for the third and youngest child in the Ernst family, Lisabet. Verena and I were practically joined at the hip, spending our days running together through the garden, while Lisabet ran after, never quite catching us. We used to tease her, often winding her up like only brothers and sisters can. This all taught Lisabet to stick up for herself – and even today, she is still the most rebellious out of the three of us.

Given my inability to talk properly, my parents were already worried that something was not quite right. They thought that the best thing for me would be to spend time with other children, so they sent me on a three-week summer camp in Aegeri in central Switzerland. This did little to help, however. I was massively homesick and spent a lot of the time in tears; my inability to talk to other children became glaringly obvious and for the first time, I felt the pain of being alone, excluded by the children at the camp from their games. I was overjoyed when my parents came to take me home. But it was then that I started to wet the bed. Finding out about the plastic sheets on my bed, the other children would tease me mercilessly. I was so ashamed that I just wished the earth would swallow me up.

I always had the feeling that my father cared more about his family's lineage than he did about me. I am convinced he thought I was mentally disabled because I could not speak properly. He studied in Munich and after the war, was involved in reconstructing Strasbourg, before moving back to Winterthur, where he soon joined the civil service. He became a professor at the School of Engineering and while he did not build many houses, he sat on many building committees. My father was strict and conservative, but as a lecturer was popular with his students, some of whom even called him "papa". His role model was his cousin Johann Rudolf Ernst, almost thirty years his senior and also his godfather. Johann Rudolf was by far the most successful member of the Ernst family. He was the son of Sulzer engineer and partner Rudolf Ernst-Reinhart, yet still managed to outshine even him. In 1912, he initiated the merger of the Bank in Winterthur with the Toggenburger Bank to form the Union Bank of Switzerland and until 1941, acted as the Chairman of the Board of Directors, before being named Honorary Chairman of the bank for life until his death in 1956. He also sat on the boards of numerous other major companies in the engineering, banking and insurance industries, including Sulzer, Brown Boveri, Georg Fischer, Munich Re and the Swiss National Insurance Company. Johann Rudolf Ernst was a leading business figure throughout Switzerland – many compared his influence with that of Zurich-born railroad and banking pioneer Alfred Escher.

The family home of this successful branch of the Ernst family was "Frohberg", a grandiose estate including a villa, summer houses, staff quarters and a sweeping driveway, situated just outside the old town up in the greenery. At the start of every year, both close and extended family were invited to celebrate the new year. This event was one of the social highlights in "Winterthur society". The well-heeled guests driving their cars up the broad driveway, tea and cakes being served to all, while the children played in the extensive gardens. But our family did not have a car for a long time, meaning we had to make our way to Frohberg on foot, envious of the brand-new cars speeding past us. We had the impression that my father always felt a little inferior to the more prosperous Frohberg branch of the Ernst family and always went out of his way to prove that his family, with its knowledge of the art world, was just as worthy.

My mother, Irma Brunner, came from a more modest background. She was the daughter of primary school teacher Heinrich Brunner, while her mother, Bertha Bauer, came from a down-to-earth family of farmers and innkeepers in Ellikon an der Thur, a small village in the countryside outside Winterthur. My mother completed an office apprenticeship and before marrying, worked in the office of the Volkart trading company in Winterthur.

She was extremely proud to marry into the prestigious Ernst family, much later telling me that it was my father's house that she first fell in love with. My grandfather commissioned the construction of this elegant brick stone villa on Gottfried-Keller-Strasse just before the turn of the century in 1898. My mother always used to pass it on her way to school and she was clearly impressed with this noble urban residence and its large garden even then. She told me that she was particularly drawn to the false art-nouveau-style window painted onto the side of the house overlooking the street, featuring ivy cascading from it as if in a fairy tale. "Oh, if only one day I could live in a house like that, with such beautiful windows," she thought. She would later go on to meet my father – and her dream became true.

My mother and father married on 16 August 1932 in the church in the village of Kyburg. By this time, both my father's parents had passed away, but for my mother's parents – grandma and grandfather Brunner – this was a big moment. They were proud that their daughter was marrying so well. But for my mother, marrying into Winterthur high society was a double-edged sword. She experienced a real culture shock and had to get used to a completely different way of life. When her dream became true and she moved into the "wonderful" villa, she was overawed and extremely shy. My mother later told me that she barely trusted herself to open a cupboard, that all she had brought to the marriage was a desk and some bedlinen – everything else was already there ready and waiting. She encountered various unspoken expectations and an etiquette that had to be followed to the very letter, ranging from art, how to sit in a chair and set a table correctly, through to engaging in proper conversation. She simply did not know how to talk to this upper-class "nobleman", who was 17 years her senior. Almost a year to the day after the wedding, I was born. At last my father had an heir, something he ascribed to fate and for which he was eternally thankful.

On August 16, 1932, Irma Brunner and Robert Ernst celebrated their wedding at Kyburg Castle. Irma Brunner was 23 years old at the time, her husband already 40 years old.

On the day I was born, my father could have had little idea just how much his son would disappoint him. He was 41 years old and performing military service in the Jura area when my mother called him to say that she was about to give birth. My mother described in her diary how he took the first available train and made it back just in time to be holding her hand during the birth.

The Ernst family on the day of mobilization for the Second World War. Father and mother Ernst with the children Richard, Verena and Lisabet (from left to right).

Having a first-born son was a small triumph for my father, but also a responsibility. His three brothers were less successful in this regard. His brother Karl died in a ski accident aged 35 and unmarried; Gottfried "only" had four daughters, who together ran a prestigious bedlinen shop in Frauenfeld; and while Franz married well – to Lily Rittmeyer, daughter of the famous architect Robert Rittmeyer, who designed the Winterthur Museum of Art and many other glorious villas in the town – and they had three sons, Lily died at an early age. My father's two sisters suffered a wretched fate. Aunt Emma Lilli did not marry and when she later fell down her attic staircase and broke her foot, she took to her bed and never got up again. Aunt Elsbeth also had an unhappy life: She married a priest, who went on to amass significant debts, which our family then had to repay. It was for other reasons that my mother's side of the family were not expected to contribute; from the outset, her family failed to meet the "demands of Winterthur high society". This meant that all responsibility sat firmly on my father's shoulders – and me as his young heir.

My father focused all his ambition on his military career. His three older brothers all achieved the rank of lieutenant colonel, but he outshone them all by becoming a colonel in the engineering corps. I remember how on the day he was mobilized in September 1939, he proudly placed his helmet on my head and placed his sword in my hand. It is no surprise that later that year I was walking around primary school claiming I had the same rank as my father!

Richard Ernst at the age of about seven.

Christmas in the family circle, around 1946. Front row (from left to right): uncle Erwin Brunner, Marina Brunner-Sulzer, grandpa Brunner, Evi Brunner, grandma Brunner, Richard Ernst, mother Irma Ernst-Brunner, Georg Brunner. Back row (from left to right): Verena Ernst, Lisabet Ernst, father Robert Ernst.

My father was frequently absent during the Second World War, away on duty on the western front in Jura. In the meantime, my mother took over duties at home. She continued the strict, traditional upbringing that my father insisted on, all in the name of doing the right thing by him. On the occasions he was there, he would work in his office behind closed doors. This meant that we had to be completely silent so as not to disturb him. Meanwhile, my mother would be elsewhere, working hard to keep the large household running smoothly, while I would be in the lounge, sitting waiting in my playpen. While there was a good amount of domestic assistance – a maid, a gardener, a laundress, a seamstress – my mother always seemed to be busy somewhere or another, either giving out new instructions to the staff or because she had to do something herself.

I also have fond memories of family life at the time: In the evenings our mother would read to us, simultaneously translating the words on the pages into Swiss-German so we could understand. I can trace my love of literature back to her. Behind his strict facade, my father also had a caring side. To mark my birth, he bought a model railway and set it up in our attic. It was a Märklin steam train, 1 gauge, with locomotives as long as your arm! He assumed I would play with it and when older, I would take responsibility for it, but model railways did not really interest me.

I always felt that I failed to meet my father's expectations. Even when many years later I sold the model railway, I felt guilty that I let it go for far less than it was worth. Perhaps this was even an act of delayed gratification, as not once did a word of praise or encouragement pass my father's lips.

Throughout my childhood, I often felt lonely and unhappy. My sister Verena later told me that she barely saw me laugh, other than when we were playing around together. We did not have many friends as we grew up in an area where there were few children we could play with. I had a favorite recurring dream, in which I felt carefree and happy: I would be floating in the air, fifty meters up, looking across the landscape, free from all the pressures I felt on the ground, free from my own body, free from the expectations of others. When this dream came to me, it was like a burden lifted.

Later on, when I went to school, my home life got somewhat better; friends would come to play with me at home, and I became more adventurous – most likely I was looking for recognition and love, something that I never felt from my parents. I once climbed onto the window ledge of the third floor and swung across to the next window, easily ten meters above the ground. Verena had to keep watch, making sure that nobody saw what we were doing. Unfortunately my mother had eyes everywhere and caught me in the act – she was shocked. She gave me a proper telling off, but it made no difference – it just made me more careful not to get caught the next time. I used to balance dangerously above the steep roof of our house, which would be extremely slippery when it got wet. But not once did I fall down.

This just made me bolder. You could say I became a "bad boy". I was once so angry about receiving a bad grade that I intentionally ruined a teacher's mark book. Even when at grammar school, I was known to fool around. My father once received a letter from the headmaster: "Your son Richard received a detention for poor behavior," began the reprimand from July 1948. "It appears that the boy has a negative influence on his classmates ... and must report for detention this coming Saturday, 10 July at 2 p.m. in the headmaster's office." My "crime"? A friend, Werner Hablützel, and I once skipped class on a sunny afternoon and instead of being in the classroom, diligently following the curriculum, we decided to look for ways to distract people by using mirrors or pieces of glass to reflect the sun into people eyes. But instead of doing this at home or in the garden, we took up a strategic position not far from the classroom where our fellow pupils were sitting, with a direct line of sight. So we had great fun directing the beams of light onto our friends' desks and into their faces – a distraction that those in charge did not find amusing.

Adventures in the world of chemistry

I spent the first thirty years of my life in our family home at number 67, Gottfried-Keller-Strasse. It is here I passed through childhood, with all its ups and downs. Is here also that I also experienced my first scientific epiphany, but more of that later. The house stood in a large garden with mature fruit trees, in the middle of a leafy neighborhood not far from Winterthur station. While it was not a huge, free-standing villa like "Frohberg", it was still a three-story, red-brick villa, with a steep sloping roof punctuated with a number of bay windows. It was a wonderful house that inspired my imagination. It had beautiful high ceilings with ornamental plaster features and in the elegant stairwell there were lights and windows decorated in the art nouveau style. "Anna lives in a castle", my daughter Anna's friends used to joke when I lived there for a while with my own family. As a key example of late 19th century architecture, the building is now even subject to a preservation order.

As children, we enjoyed great freedom in the adventurous house and its large garden. Everybody had their own room, each with a small wash basin and tap, which saw me get up to all sorts of flood-related mischief. The villa had brushed concrete floors and an airy cellar; in fact, it was much too big for our small family, which is why we always let out the middle floor. But the dark attic and its various compartments belonged to us children. For us, it was a sort of ghost castle, full of secrets and magical adventures. In front of a gothic-style bay window stood a large wooden cross, which was supposed to serve as a reminder to lead a godly life. However, it actually served as gun stand, where my father and his three brothers kept their ancient service rifles, always ready for the next call to arms.

Richard Ernst's birthplace, a city villa from the Wilhelminian period, built by grandfather Walter Ernst in 1898. In the basement, Richard Ernst conducted daring chemical experiments as a boy.

My father had set up a workshop in the basement. I slowly took over this space for myself, at first hesitantly, but then with a passion. I would soon slip away into the cellar whenever the chance arose, to busy myself with one thing or another. I didn't have to put up with being controlled, dependent on someone or corrected, especially not by my father. The workshop gradually became a refuge for those occasions on which I had had enough of the "normal" people upstairs. Here I was able to do – or indeed not do – anything I wanted.

When I wasn't downstairs, I lay on the bed in my room, devouring the works of Karl May as well as adventurous tales of travels to distant lands in Africa, Asia and South America. This faraway world was a source of endless fascination and awoke a sense of yearning and youthful curiosity within me.

Our second world was the garden, which stretched back to the railway platforms leading to eastern Switzerland. Anyone traveling to St. Gallen, Romanshorn or Stein am Rhein can even today take a peak over the fence as they speed past. In those days, however, trains from all over eastern Switzerland rolled slowly by, pulled along by their heavy locomotives. Towards the end of the Second World War, trains carrying wounded soldiers from the German battlefields traveled from Stein am Rhein to Winterthur, where the soldiers would recover. They had an other-worldly feel to them; they sometimes stopped just short of the station and we were able to see straight into the soldiers' eyes. We would wave to the injured or on occasions, sneak through the fence separating our garden from the train lines and throw the men oranges, cigarettes or chocolate through the open windows, which they gratefully received.

The Second World War passed us children by like a haze, as was the case for most children who grew up in our country, which was thankfully spared the ravages of war. The most dramatic consequences of the war for us was, as mentioned, the fact that our fathers were absent most the time. As part of Switzerland's drive to increase domestic food production due to the reduction in imports (known as the "Anbauschlacht"), my family had grown a vegetable patch in our garden and we kept rabbits, which we could play with – at least until it was Sunday roast time. That is just how things were in those days, but it still came as a shock for us children when we saw our floppy-eared friend "Fritzli" appear in the roasting pan. And we all used to listen to Radio Beromünster, a station that carried the latest reports of how the war was going – we would crowd around the radio and listen as soon as our father got home. There was never any doubt in our family as to who the "bad guys" were in this war, that is to say our northern neighbors. At the time, we did not realize the reasons for this or the importance of events.

As a boy, I was much more interested in the adventurous nature of what was going on rather than the political aspects. For example, when I used to peer through the skylights with my binoculars to spot the GI bombers thundering above us on their way to Germany, a shiver would go down my spine. This was probably all the more exciting as our parents had strictly forbidden us from doing so given the blackout rules in place. How should I have known that these airplanes posed a real danger? It was only later I found out that Schaffhausen, a town only twenty or so kilometers away from

Winterthur, had been bombed by the Americans in 1944, leaving 37 dead and hundreds injured.

It must have only been a few months after the war ended, when I was 13, that I made a fateful discovery. It was on one of my forays into the attic that I found a crate of glass bottles full of various chemicals. I later found out that these belonged to my Uncle Karl, who was an engineer, but had a keen interest in photography and chemistry. Uncle Karl died in a skiing accident long before I was born, but some of his belongings remained forgotten for many years in our large attic, that is until I stumbled upon them.

One after the other, I carried the bottles down into the workshop in the cellar and built my own little chemistry lab. Then I began to experiment. I soon also found a Bunsen burner, which I connected to the gas line feeding our lights. I set about performing various basic chemistry experiments: mixing, melting, evaporating, distilling... I even did some glass-blowing. I remember there being unlabeled bottles of concentrated hydrochloric acid and other dangerous chemicals lying around in the cellar. My uncle's chemistry set was definitely not child-friendly by today's standards; there were no instructions or warnings, no reminders that experiments had to be carried out "only under adult supervision".

I gained a basic knowledge by reading the chemistry text books I found in my father's library and then later on, from the public library in Winterthur. I remember one book called "The School of Chemistry or The Principles of Chemistry: Illustrated by Simple Experiments" written by Dr. Julius Adolph Stöckhardt, "Privy Councilor to the Royal Court of Saxony and Professor at the Royal Academy of Tharandt", published in 1870 in its 16th revised edition. If I were to now nostalgically flip through its yellowed pages, I would come across chapters with titles like "Inorganic chemistry" and find mentions of "Vital forces and chemical processes" in the introduction – descriptions and explanations that sound like they were written in another era. It contained a lot of information that was outdated even then and I would subsequently have to "un-learn".

But for me, this secret world was a revelation. I had no idea that at the very same time on the other side of the world in California, a Swiss physicist named Felix Bloch was carrying out experiments that formed the basis for nuclear magnetic resonance, the subject that would later be my specialist field. Neither could I have imagined that just a few years later I would meet Bloch, who went on to receive the Nobel Prize in Chemistry in 1952, and have the chance to learn directly from him.

Having retreated into my own little world in the cellar, my initial experiments often produced unexpected results and I was constantly discovering new, amazing reactions. I felt like an explorer, navigating my way across unexplored territory, with nature's last secrets just waiting to be discovered. Spurred on by my unbridled curiosity, chemistry became a wonderful diversion of which I never tired.

I now realize that this obsession was just as much a way of fleeing my complex existence. In my desperation to gain acceptance and recognition, I had found something that gave me respect for myself. While my father did not forbid me from experimenting with chemicals in the cellar, he did not support me either. My mother, meanwhile, was a little uneasy about my hobby, but let me carry on regardless. At school, I had found a subject in which I could achieve good grades without it particularly impressing the teacher. So chemistry became the subject that belonged to me, and just me, a subject in which I could also stand out from the others. I wanted to be the only person who could do it, I wanted to be the best at it. I wanted to be something special. Neither my parents nor my classmates had the slightest idea about chemistry and there was nobody I knew who had anything to do with the subject whatsoever – this science was so alien to them that they were not impressed by any of it, but this did not bother me in the slightest. The feeling of adventure that chemistry gave me was more than enough.

On the level crossing with Igor Stravinsky

It goes without saying that as a "son of the Ernst family", I would have to learn a musical instrument. I must have been eight or nine years old when I began to learn the recorder, starting with a descant before moving on to a treble recorder. I was recently visited by someone from the Nobel Prize Museum, who asked me for these two recorders, and they are now on display in Stockholm in the "Richard R. Ernst" display case. I never actually liked the recorder that much and would have preferred to play the piano, but that did not fit into my father's strict plans. Following mandatory recorder lessons, he had in mind a specific instrument for each of his children: for me it was to be the cello, for Verena the violin, and the piano for my youngest sister Lisabet. The idea was for the three of us to perform small concerts at social events, as was befitting for a cultured family such as ours. Verena and I followed his plans without complaining, but Lisabet was less willing. She never quite mastered the piano and had difficulty putting in hours of practice. But she knew how to stick up for herself and she got her way: The experiment came to an end.

I, on the other hand, developed a real passion for music, fed by the cultural atmosphere that prevailed in my hometown. Nowadays, many Swiss people consider Winterthur to be a dull industrial town, but this ignores its other side: its museums, its art galleries, and the Musikkollegium – a symphony orchestra which I support financially to this day. It would come as no surprise if Winterthur were one day named a European Capital of Culture. How it must have been during the war, when artists and musicians came to the town, unable to perform elsewhere in Europe! Winterthur became a cultural hotbed: the town was home to a famous symphony orchestra and solo artists flocked to the auditorium in the town hall. The legendary Catalan cellist Pablo Casals was there, as was composer and conductor Igor Stravinsky, whose works include The Firebird and the Octet for wind instruments (the original manuscript can indeed still be found at the Rychenberg Foundation in Winterthur). The Romanian-Jewish pianist Clara Haskil also performed there and went on to become a Swiss citizen.

And as child, I found myself right in the middle of this amazing environment! I often saw the artists face-to-face as they walked to the Villa Reinhart after performing, where they would be meeting with patron of the arts Werner Reinhart, who sometimes offered them accommodation for the night. To get there, they had to go over the level crossing in front of our house and I would often see famous artists waiting for the barriers to raise. As a child I didn't actually know the musicians and I wasn't confident enough to approach them. Yet their mere presence and the charisma they exuded during concerts left a lasting impression on me.

Whenever I went for a recorder lesson, my teacher, Linda Bach, would admire my hands, telling me that I had the fingers of a talented cellist. I must have been flattered by her comments as I began to learn the cello aged eleven. But I was not destined to be a star pupil, no matter how much I wanted to be. I couldn't even get the strings to make the sounds I was aiming for; much to my disappointment, my manual dexterity always lagged behind my hearing. And I found the lessons to be boring; they felt much more like I was having to laboriously learn a craft and did not provide me with the feeling of elation I experiencing when listening to music.

Ultimately, I preferred music theory. In our music lessons at school, we had to practice melodic dictation – something that I was regularly the best at. I could instantly recognize the intervals and transcribe them correctly. I also loved analyzing music and composing it myself. When I was twelve, I wrote some small pieces for two string instruments, later adding a piano part. My sisters and I premiered my "works" on our parents' birthdays in front of their half-open bedroom door.

I soon wanted to start learning the piano so I could hear my compositions in multiple parts. But my father did not allow me to learn a "second" instrument; he thought it more beneficial for me to take a typing course. "You can tap keys there as well," he would tell me. I was of course extremely disappointed, but what could I do? I drew a piano keyboard on a piece of paper and practiced without sound, as it were. And as soon as my parents left the house, I would sit down behind the family piano and tried to teach myself how to play or practice my compositions.

The melodies I had captured on paper fascinated me almost more than the music itself. I used my pocket money to buy musical scores of all the famous works and then took them with me to the rehearsals for a series of concerts that we were allowed to attend for free as music students. I would sit there with my musical scores, following along with what was being played on stage, frequently discussing them with my equally enthusiastic friends. This is how I made music my own. I liked the idea that by having the scores, I somehow "owned" the music and could therefore "capture" the transcendental power of the melodies.

Classical music played an important role in my youth; it helped my "survive" my crippling self-confidence. The first job I ever aspired to was that of a musician or conductor, but I simply didn't have the talent required. I now look back and think that my obsession with musical scores somehow helped lift the lid on the secrets of chemistry. After all, in the broadest sense is chemistry not comparable to a symphony? For example, how the individual instruments – the violins, the cellos, the horns and percussion – merge into a meaningful whole, much like how elements, atoms and molecules combine to form something new. And like how the conductor has to master the individual scores, a chemist must have complete control over the substances and their properties so that instead of creating chaos, something beautiful can emerge.

My passion for classical music remains with me to this day. I still have a collection of miniature scores that I bought with my pocket money as a boy, and while at my age, my memory is not what it used to be, I can still hear the pleasant melody of a Bach solo suite or Stravinsky's Octet clearly playing through my mind.

"One day my elder brother will win the Nobel Prize"

My youthful "Sturm und Drang" years came to an end one night in a police cell. There used to be a traditional restaurant on one of Winterthur's main shopping streets, the Marktgasse, called Restaurant Walfisch. At the time, it used to regularly hold folk music concerts, while before and during the war, it was notorious as a meeting place for the *Fröntlers* – Swiss Nazi sympathizers. It was still popular following the war, mainly among right-wing, conservative circles, the "natural" enemies of us high-school students.

We would often try to disrupt their meetings, more out of a sense of youthful rebellion than political conviction. In 1951, our rebellious nature got us into real trouble. The Restaurant Walfisch was holding another political event and an older friend of ours had illegally got hold of some tear gas rounds from the army supplies. As the meeting was taking place, we threw some of them through the window. Our "enemies" sought cover and saved themselves by jumping through the doors and windows onto the Marktgasse. I can't remember whether I actually threw any of the tear gas charges, but I was definitely there. We watched events unfold with a real sense of satisfaction, but our Schadenfreude was short-lived; passers-by had reported us to the police, who quickly took us into custody and brought us to the station. When they found out that we were from prominent Winterthur families, the situation was defused. We were given a fine and needless to say, my father was livid that his son had sullied the family's good name by getting involved in something like this.

The senior class of the Kantonsschule Im Lee of teacher Hans Läuchli (4th row, 1st from right), 1951. Richard Ernst (2nd row, 3rd from left) is seated next to his classmate Werner Hablützel (1st row, 1st from left).

But the idea that I was a particularly rebellious youth was wide of the mark. For the most part, I was a quiet, fairly inconspicuous high-school student. At home I did what I was told and stoically accepted the strict relationship I had with my parents, as did my sister Verena. After all, we had everything we could ever want, except perhaps love and recognition. "Born stupid and haven't learned anything since," my father used to say to me when he was unhappy with my achievements. I suffered from crippling self-doubt and a lack of self-confidence. I was never satisfied with myself; I don't know whether this was a curse or a blessing. The word "satisfaction" just isn't in my vocabulary, even today. I often ask myself whether I even deserved the Nobel Prize. This isn't an easy thing to live with – not for me, and not for those close to me.

In those days, I was also very interested in literature. I was particularly fascinated by the great Russian novels, for example: I devoured Dostoevsky's "The Idiot" and "The Brothers Karamazov. The works of Leo Tolstoy, Alexander Pushkin and Boris Pasternak made a lasting impression on me and I also loved the novels of Swedish author August Strindberg. I was particularly struck by Hermann Hesse's poem "In the Fog" the first time I read it. It seemed to be a perfect expression of how I felt:

Strange, to wander in the fog.
Each bush and stone stands alone,
No tree sees the next one,
Each is alone.
My world was full of friends
When my life was filled with light,
Now as the fog descends
None is still to be seen.
Truly there is no wise man
Who does not know the dark
Which quietly and inescapably
Separates him from everything else.
Strange, to wander in the fog.
To live is to be alone.
No man knows the next man,
Each is alone.

I remember my years at high school as a period of emotional and intellectual anguish. The things we would learn were pointless to me. Languages, in particular, were a struggle; my grades in English, German and French were middle of the road at best. I detested French, and couldn't string together a sentence in Latin. In my favorite subjects of chemistry and physics, however, I was already way ahead of even the teachers, meaning I had to change high schools, moving to a more maths- and physics-oriented school. In languages, I simply couldn't remember any of the words. By contrast, even the most complex chemical and mathematical formulas were readily committed to memory.

I was diagnosed as being dyslexic, which at the time was the condition ascribed to children who had trouble learning languages, who couldn't form sentences easily, who mixed up their letters. My father was also dyslexic; whenever he had to write a letter or report, he always had my mother proofread it before sending it off. Nowadays, I would most likely be diagnosed as having Asperger syndrome, which is a mild form of autism. My wife Magdalena, who for many years was a primary school teacher, read a great deal about the condition and says that I exhibit almost all of the conditions typical of someone with Asperger's.

It is not just foreign languages I struggle with, it is talking itself. I simply didn't like having conversations, particularly in front of others. I had no wit or repartee and so would often say nothing at all. My fellow students would think I was eccentric, arrogant possibly, a mummy's boy from a good family. When studying at university, I turned down the chance to be a research assistant, mainly because the thought of having to speak in front of a room of students was too much to bear. My colleagues thought that as a member of a wealthy family, I simply didn't feel the need to take on the position, but this was not the case. I was a loner because I couldn't be anything else, not because I wanted to be.

The culture of communication in our family was no help. There simply wasn't one. When we sat down to eat together, the children had to stay quiet. The only sound to be heard was the news on Radio Beromünster. If there was every any argument amongst the children or in the family, my father ended it with the words: "People like us do not argue – that is something the working class do." Our family life was smothered by a fog of unspoken emotions and silence. I remember once having guests at home. We were sat in the living room and nobody had any idea what to talk about, so my father took a book from the shelf and started reading aloud. After that, we didn't have many visitors. It was mostly just us. The only regular guests were my grandparents, from my mother's side of the family, who came over every Sunday afternoon for tea and cakes. But instead of using the time to discuss important, or even minor family matters, my father would disappear into his office.

I resigned myself to this atmosphere. My sister Verena recently told me that I most likely suffered from the fact we didn't talk about anything. Perhaps I adapted to the situation, but perhaps the situation adapted to me. I would also hide away in my bedroom, busying myself with things I actually enjoyed. I immersed myself in my books and worked on my experiments. It was here that I developed an irrepressible desire to finish anything that I started. There was never any question of abandoning anything prematurely or leaving something unfinished. My sisters thought me to be somewhat of a swot, a bookworm hiding away learning things alone. "One day you'll win the Nobel Prize!" they would joke on more than one occasion, more teasing me than actually predicting the future.

This unyielding resolve, this inner urge to prove myself and to achieve my goals at any price, my ability to focus single-mindedly on my lab work, all of this is what played no small part in my life to date. Many years later, a fellow researcher told me that he had never met anyone who could work so hard and tirelessly. But my main aim at the time was to make it more or less unscathed through the battlefield of my teenage emotions. The vibrant world out there, the temptations of youth, all of those things held less appeal to me than perhaps they should have.

By the time I left high school, I had got any problems I had in certain subjects properly under control, meaning I passed all my exams, and scored straight A's in maths, physics and chemistry. The path was now clear for me to study chemistry at university. However, it soon turned out that embarking upon my degree did not free me from my demons as I had hoped.

Undergraduate Degree and Dissertation at ETH Zurich, 1952–1962

Alma mater

In September 1952, I passed my high-school-leaving certificate and just a few weeks later, embarked upon my degree in chemistry at ETH Zurich. But it was a year before this that I became sure of the path I wished to tread. Thanks to my father's contacts, I was able to get an internship in the laboratory at the Hovag wood saccharification plant in Ems, which would later become Ems-Chemie. This company in the mountainous canton of Graubünden used to convert vast quantities of local waste wood into ethyl alcohol, which during the Second World War was added to petrol – then a scarce commodity – as a fuel additive. In the post-war years, Hovag was a prestigious company, because in Swiss people's eyes, the saccharification of wood was seen as a key element in what was termed as the country's "intellectual national defense", as it helped maintain Switzerland's independence.

During the Second World War, Hovag covered around thirty percent of Swiss fuel requirements and received generous support from the Federal government in return. The plant in Domat-Ems, at the time still located some way outside the village in the countryside, quickly became the largest employer in the canton of Graubünden. After the war, the government subsidies began to dry up and the company diversified into the production of various chemicals and became Ems-Chemie, which has been owned by the Blocher family since 1973. The Swiss politician and pastor's son Christoph Blocher gained wealth and power on the back of Ems-Chemie. He later went on to mold the Swiss People's Party (the *Schweizerische Volkspartei,* SVP) – which originally represented mainly farmers and small business owners – into a staunchly right-wing, populist organization, in the process enjoying major success at the ballot box and in 2003, even being elected to the Swiss government, the Federal Council. After I was awarded the Nobel Prize in 1991, Christoph Blocher heard that I also had a past link with Ems and he invited my wife and I to attend the Albisgüetli-Tagung, the SVP's party conference. Politically speaking, I have never been on the same wavelength as Christoph Blocher, but we still accepted the invitation to the conference. Happily we were seated next to the former Federal Councilor Leon Schlumpf, the father of Eveline Widmer-Schlumpf, who herself later went on to become a Federal Councilor and in doing so, ousted Christoph Blocher from government.

In the 1950s, an internship at Hovag was the perfect opportunity for an aspiring chemistry student to gain practical experience in a laboratory. During the week, I would carry out water analysis in what was then a state-of-the-art industrial laboratory. Highly purified water was important for the production of polymer compounds. I didn't receive a normal salary for my work – just a small "grant" of fifty or so francs a day – but I enjoyed working on a task that then also had a practical application. I was also able to draw on my childhood experiences of experimenting in my basement back on Gottfried-Keller-Strasse.

I really enjoyed my time in Graubünden: In my free time I would go hiking with my schoolmate from Winterthur Walter Jung, or we would cycle up and down the many mountain passes the area had to offer. Walter was a distant relative of the psychiatrist Carl Gustav Jung, a fact that would later play a major role in my life. But first of all, I trod the more or less straightforward path of a chemistry student.

After high school, I registered at the renowned Swiss Federal Institute of Technology in Zurich (ETH), as had always been the plan. Full of enthusiasm and with high expectations, I began my degree in autumn 1952. At the time, an undergraduate degree was more a continuation of high school, albeit at a higher level; there was a strict timetable containing many lectures and a lot of lab work. The students (almost all of whom were male) were split across year-long courses and had little freedom in choosing lecturers. Those wanting to qualify as a certified "ETH Chemical Engineer" after four years had to first pass two intermediate diplomas and one diploma exam, the latter requiring theses to be submitted in four subjects: organic, inorganic, physical and technical chemistry. I was much too hard-working to think about skipping lectures and I diligently attended them all. I was not even able to find a great deal of time for my beloved music.

I have few real memories of my time spent studying – it was mostly just bland, day-to-day life. I would commute every day by train from Winterthur to Zurich. A colleague, also from Winterthur, who had the same commute as I did, would always take the slower local train passing via Kloten so that he had enough time to read the local newspaper, NZZ, from cover to cover; meanwhile I would take the express train. I didn't really experience much of Zurich as a city. Each day I would walk straight from the station up to ETH, usually to the old chemistry building. This is now a listed building, and rightly so; it is within this late-nineteenth-century brick building that Swiss scientific history was made. No less than seven Nobel Prize winners studied, researched, or lectured here, myself included.

Something I have very vivid memories of are the train journeys home in the evening, when I would usually have a compartment all to myself. I was, of course, a bit of a loner in my youth, but the reason for being left alone in the train was far more mundane – as any chemistry student can confirm. Every afternoon, we would have practical exercises in the laboratories. Because we had spent hours using odorous solvents and chemicals such as ammonia or acetic acid, our clothes had become impregnated with such foul odors that we would regularly be followed around by a disgusting smell on our way home. It was clearly so pungent that all the other train passengers would move away, holding their noses, leaving the chemistry student Richard Ernst from Winterthur all alone, literally repelling women without trying!

In the first semester, I took the modules "Basics of inorganic chemistry and analysis", "Basics of organic chemistry" and "Differential and integral calculus". For my second intermediate diploma, I had to take exams covering the "Basics of physical chemistry and electrochemistry" as well as physics and mineralogy. Overall, I achieved good grades, but I was clearly most interested in physical chemistry, where I achieved the highest grades possible from the outset.

Opting for physical chemistry

Outsiders often have difficulty understanding just why I am so fascinated by physical chemistry. While everyone has an idea about classical chemistry – where scientists are in the lab, mixing substances in test tubes, which bubble away and produce loud reactions and, ideally, something new; a field which is seen by many even as a creative art – physical chemistry, with its focus on math, is a mystery to many people. Also referred to as "theoretical chemistry", it aims to explain the interactions of elements using the laws of physics and mathematical calculations. Its foundations lie in the laws of thermodynamics, the gas laws and, following the quantum revolution, primarily the laws of quantum mechanics, which have given us a far deeper understanding of chemistry over the last century.

In the 1950s, however, physical chemistry at ETH Zurich was still a purely academic field, whereas classical chemistry and therefore the Laboratory of Organic Chemistry, in particular, were flourishing two highly decorated professors, the Croatian-Swiss chemist Leopold Ružička and the Sarajevo-born Vladimir Prelog, who Ružička helped flee to Switzerland during the Second World War, taught here. These two giants in their field were not only first-class researchers, but fascinating men as well. I got to know Ružička, who had already won the Nobel Prize in 1939 for his research into natural substances, in his final years as a lecturer at the university. He was an interesting man in many ways and also a passionate art enthusiast and collector, as I went on to be. Vladimir Prelog, twenty years Ružička's junior, was already at the height of his powers, going on to receive the Nobel Prize himself in 1975. As a professor, Prelog was extremely modest and approachable, far from the aloof scientist one might come to imagine. He was always open to suggestions and frequently asked his PhD students about their findings. He was also responsible for being the first person at ETH to bring the American team model to the 1950s' Laboratory of Organic Chemistry. Within his research group, he viewed himself more as a "village elder", rather than being the absolute ruler of an institute, as was previously the norm at German-speaking universities.

I would therefore have had a million reasons to head into the area of organic chemistry, yet I felt that I wouldn't find the answers there that I was looking for. By contrast, physical chemistry was shaped by the exciting scientific revolution that was quantum mechanics. I was most fascinated by atomic structures and the related theory. While this was difficult to understand, it spurred me on instead of putting me off. It is a fascinating construct of ideas, which in principle is logical and comprehensible, yet whose radical core tenets about the essence of matter, but also about perception and truth, radically changed the way we viewed the world.

However, there was little sense of this evident in early physical chemistry lectures at ETH, which instead focused on topics related to classical thermodynamics and electrochemistry. It was as if the quantum revolution in chemistry had completely passed Zurich by, despite the city having played a key role before the war in this area. In the golden era of the 1920s, for example, when Europe was experiencing not only a societal, but also scientific boom, it was during his time as a professor at the University of Zurich that Austrian Physicist Erwin Schrödinger developed his famous wave equation, which would have a radical impact on chemistry. And when I was at ETH in the 1950s, Wolfgang Pauli – another giant of the quantum revolution – was still teaching there. I have faint memories of attending his theoretical physics lectures, which at the time were almost akin to an act of public worship. This master of quantum physics, with his imposing physique, stood at the front of the room at his blackboard, spending minutes in deep thought, moving his head back and forth, looking at the formulas he had covered the board in. He would then suddenly scribble a correction or develop another formula, all without noticing that behind him were sitting dozens of inquisitive students, all trying to follow his pattern of thoughts in dazed confusion.

Who knows, perhaps this patchy way of learning even helped me in the long run? There was nothing else to do apart from head to the library and further my knowledge myself, just as I had done for all those years before at home. I just wanted to know more! But there was nobody to help me or even give me a hint as to which books I could read to find the answers to all the pressing questions I had. So I had to find this out for myself as well. My favorite resource at this time was a text book by Samuel Glasstone called "Theoretical chemistry", which taught me the basics of quantum mechanics, spectroscopy, statistical mechanics and statistical thermodynamics. This is how I navigated the minefield of learning the mathematical and physical principles of my chosen specialism.

One exception among the dull lectures at ETH were the voluntary physical chemistry courses organized by Hans Heinrich Günthard. Günthard was a young, enthusiastic lecturer with an already amazing career. He had originally completed an apprenticeship as an electrician and then studied chemistry at the School of Engineering in Winterthur. During and after the Second World War, he worked for a number of years in industry, specializing in military radio technology. It was only then that he embarked upon a degree at ETH, where he completed a chemistry and physics degree in parallel. He later became a research assistant at the Laboratory of Organic Chemistry under Leopold Ružička and in 1949, completed his doctorate in spectroscopy. By 1951, he had already become an outside lecturer in physical chemistry and in 1952, he was appointed associate professor. As a student, I was really impressed by his lectures, finding them clear and illuminating. I had soon decided that I would like to complete my doctoral thesis – the logical next step in a research career – under Günthard.

A liberating tragedy

In the summer holidays of 1955, something happened that derailed my life somewhat. I was away with my younger sister Lisabet; while we had frequently argued as children, we had since grown up and now got on well. She was at high school and I would often sit with her, patiently helping her with her homework. That summer, we had taken our folding canoe and tent to spend a few days meandering up and down the Danube. We had already reached Austria when we came across another boat, which – like ours – was flying the Swiss flag. In those days, summer tourism was yet to take off, so it was completely normal to greet your fellow countrymen when you saw them abroad. But these tourists were waving wildly at us, shouting across to ask whether we were the children from the Ernst family in Winterthur. We were amazed that they knew us by name. Once we had confirmed it was us, they gave us the sad news: We should return home immediately; there had been a death in the family and people were looking for us all over Europe – they had heard it on the radio. At the time, it was common to search for people over the radio; after all, there were not that many stations.

We knew that our parents were spending time at a spa in Bad Neuheim. My father was already in poor health, having previously suffered a heart attack. I remember how he always used to walk extremely slowly, often stopping – in his words – to admire the flowers by the side of the path. He was most likely trying to conceal his heart problems. Before the summer holidays, his doctor had prescribed a stay at a therapeutic spa, so my parents had gone to Bad Neuheim. But when walking in the park one day, he suffered another heart attack, which on this occasion proved fatal.

When my sister and I heard the news on the Danube near Linz, we straightaway headed to the nearest village, where we packed our things and took the train home. I no longer remember how I felt at the time. Perhaps we were holding out the hope that it was our grandfather on our mother's side, who was already very old and ill. The tourists who broke the news to us couldn't provide any further details other than that a family member had passed away. But when we arrived back in Winterthur, my father was already in the coffin. It was almost surreal to see him lying there, still sporting his impressive mustache, which my mother would have loved to comb for him at this moment.

My father died on 2 August 1995 in Bad Neuheim aged just 63. Of course it came as a shock, but my relationship with him had never been particularly close. As such, I experienced a flood of mixed feelings following his death. On the one hand, I felt I had been freed from the burden of my father's expectations, while on the other, I felt an increased weight on my shoulders. I was suddenly responsible for the whole family, as the patriarchal family model dictated in the 1950s. I was the only son, and the first-born child as well, so I automatically had to step up. Out of a sense of obligation and wanting to always do the right thing, this was a duty which I did not intend to shirk. I suddenly had to take care of everything that needed doing in and around the house. To a certain extent, I also become a substitute partner for my relatively young mother in everyday matters. She suffered greatly following the death of her husband, despite him never being the perfect family man when he was alive.

My emotional ambivalence towards my father's death was most clearly reflected in my military career. The military represented one of the most important facets of my father's self-image, possibly more important than his family. Like his relatives before him, he had wanted to have a career in this area, making it as far up the ranks as a colonel in the engineering corps. He had tried to pass on this ambition to his son; on the day he was mobilized in 1939, he placed his helmet on my head and his sword in my hand. During my rebellious youth, however, I steadfastly refused to live up to his expectations. Perhaps I only imagined this, but in any case, I swore not to follow the same path as my father. I once said that if anyone tried to put a weapon in my hand, I'd throw it away. I was fundamentally opposed to the military – I had the blood of a pacifist coursing through my veins. But after my father's death, I somehow felt obliged to carry on his legacy. So I ended up following in his footsteps, embarking upon a career as an officer, mainly out of a sense of remorse.

After qualifying as a certified "ETH Chemical Engineer" in August 1956, I spent more than a year at officer training school, becoming a lieutenant in the infantry. As a chemist, I was made an "AC-Offizier", a specialist focusing on protecting the country against nuclear and chemical weapons. This earned us the nickname "gas uncles". My memories of this time, however, are not overly positive – a sense of boredom mixed with an aversion towards the stupidity of the military in general, with its endless marches, to be completed carrying all manner of heavy equipment. Looking back, it is clear that this extraordinary situation painfully laid bare my inner turmoil. On the one hand, I worked hard, I became a sports officer in an infantry company and had to take part in numerous competitions. Thanks to my thoughtful and ponderous nature, I even enjoyed great success as a pistol marksman. On the other hand, I endured many a sleepless night, with nightmares a common occurrence. The teasing I had suffered as a child for wetting my bed reemerged from my sub-conscious and my father regularly visited me in my dreams. I felt like I was walking a tightrope, either "on top of the world, or in the depths of despair", possibly a manifestation of a deep-seated schizophrenia, a character trait that is likely to accompany me throughout my life. But in the military, I also learned not to take myself too seriously, a quality that I still display to this day.

Richard Ernst with sisters Lisabet (left) and Verena, in 1957, during his training as a lieutenant in the infantry.

What is NMR?

I drafted my first scientific publication before even beginning my doctoral thesis, while I was completing an industrial internship at the Ciba chemical company in Basel. I worked there for six weeks in spring 1955 – before my father's death – in the area of chemical dyes. The work had nothing to do with my later specialist field, nuclear magnetic resonance. According to the internship certificate Ciba issued me, I was involved with "preparatory and physico-chemical work". It was pure laboratory chemistry.

The fact that my internship resulted in a paper being published in what was then one of Switzerland's most prestigious scientific journals *Helvetica Chimia Acta* was of course a marvelous and, for a mere student, unexpected success. As was common at the time, the paper was published in German. I was named as the first author and my supervisor was Heinrich Zollinger, then the group manager in "my" Ciba lab.

I was really lucky to have had Heinrich Zollinger as my first boss. I really valued spending time with him and I have fond memories of the many discussions we had about the subject I liked most of all: Chemistry. He later became a dye chemist at ETH Zurich and in the 1970s, was actually Rector of the university. For a number of years, he chaired the Foundation Board of the Swiss National Science Foundation and was awarded the "Order of the Rising Sun" by the Japanese government. In an ETH anniversary publication in 2005, Heinrich Zollinger mentioned our publication together, flattering me in the process. He died later that year aged 86.

The reference Ciba gave me after my internship confirmed that I had performed all the tasks assigned to me to their full satisfaction and said that I had proven myself to be a "capable prospective chemist". "With regards to his character, Mr. Ernst made an excellent impression," it went on; this last sentence still brings a smile to my face even today.

How on earth could the chemists at Ciba judge this in just six weeks? I assume that I simply came across as conformist, eager, friendly and shy in everything I did. In no way did the problems I had approaching others or speaking in front of people abate following puberty; these traits still followed me about. I absolutely hated being in the spotlight; I much preferred being able to work away on my own. Neither did I even like expressing myself in words; I was horrified by the idea of talking in front of other people and would regularly panic at the thought – not exactly the qualities required of somebody aiming for a career in the cut-throat world of academia. There was only one solution to this problem: I had to be better than the others, I had to be different from them.

Especially after winning the Nobel Prize, I was often asked to explain how I came to specialize in nuclear magnetic resonance and why I specifically chose this area. When this happened, I would always emphasize how fascinated I had been by Hans Heinrich Günthard's lectures. But this was only part of the truth. I suspect that I subconsciously chose this subject so as to prove to myself and, above all, others that I could be successful in such a difficult field, but also to help forget the problems I had with my own being. I would flirt with failure, only then to be able to reassure myself that I wasn't to blame for not succeeding – it was down to the difficulty of the task at hand. It was a case of psychological projection: I would blame the environment I was in for any perceived lack of success on my part. I was convinced that everyone overestimated my abilities and that I was nowhere near as good as I pretended to be.

The fact that Hans Heinrich Günthard took me on as a member of his staff despite my doubts must have been a favorable turn of fate. The way I was brought up meant that giving up or backing down wasn't an option. While studying for my intermediate diploma at ETH, I didn't come across the term "nuclear magnetic resonance" once. On beginning my doctoral thesis, I knew little more than the absolute basics: That is to say that molecules consist of nuclei and electrons and that certain nuclei are not simply rigid spheres, but are magnetically active. If you expose them to a magnetic field, they orient themselves like needles of a compass and begin to rotate around their own axes. The speed at which they rotate, that is to say their frequency, depends on both the strength of the magnetic field, but also on the specific properties of the nucleus. In chemistry, it is said that nuclei have "spin". If these nuclear spins are then exposed to radio waves of a certain frequency – the resonance frequency – they absorb this radiation and then respond by producing a signal that provides a lot of information about the nucleus. There is a famous picture of Hans Heinrich Günthard demonstrating nuclear spin in his own charming way, gently swinging his hips to keep a hula hoop rotating around his midriff.

These findings were still quite new when I began my dissertation in 1955, however. It was not until 1939 that the American Isidor Rabi and 1946 that his compatriot Edward Purcell and the Swiss Felix Bloch simultaneously succeeded in measuring the magnetic moment and resonance frequency of certain nuclei, for which both men would go on to win a Nobel Prize in Physics. At the time, however, this nuclear magnetic resonance method (also known as NMR spectroscopy) was simply basic research, the main aim of which was to better understand the physical structure of matter.

Hans Heinrich Günthard demonstrates the basic property of the nuclear spin with a hula hoop, around 1958.

There was still no indication that this method could be used to precisely identify not just the composition of chemical substances, but also their structure and appearance, and that it would go on to revolutionize medical imaging. Scientists still lacked the technical knowledge required for nuclear magnetic resonance to have a practical application. War-time research laid the foundations for this step, primarily in the US and Great Britain. Quite early on in the Second World War, the Allies pushed forward with the development of high-performance radar equipment in order to gain the upper hand in the U-boat war with Germany. In 1940, the US National Defense Research Committee set up what was known as the "Radiation Laboratory", or "Rad Lab", at the renowned Massachusetts Institute of Technology (MIT) near Boston. The Rad Lab attracted the best scientists, who were tasked with developing radar equipment that could be used to settle the U-boat war in the Atlantic in America's favor. Bound to strict secrecy, the researchers and technicians learned about the practical applications of electromagnetic waves. They developed vacuum tubes, built pre-amplifiers and amplifiers, they invented transmitters and receivers that could be used to send and receive radio waves of certain frequencies and energies. These findings made their way into chemistry, where huge advances were made in the development of all types of spectrometers, from infrared and microwaves, through to UV spectrometry. The foundations for the dawning computer age were also laid during this period. The veil of secrecy was lifted starting in autumn 1944. It was the intention of the then Director of Research at the Rad Lab – no other than the man who discovered NMR, Nobel Prize winner Isidor Rabi – for the extensive research findings and technical knowledge to in future be made available to all for peaceful purposes. Rabi authored a series of publications, which were very well received by experts in the field: the books of the "Radiation Laboratory series", also known as the "Red Books".

And so with a delay of several years, this scientific knowledge crossed the Atlantic to reach post-war Europe, and indeed Switzerland. In 1949, Hans Staub – a former employee of Felix Bloch – returned from California to the University of Zurich, where he began to build on the existing findings. At ETH Zurich, it was primarily Hans Heinrich Günthard who had been closely following these interesting developments and who, at the start of the 1950s, started to construct a spectrometer – at first in collaboration with

Staub. Rabi published 28 bulky volumes of the Red Books in quick succession, with all of them becoming standard reading in Günthard's department. Günthard had recognized the sign of the times and was committed to ensuring that the new spectroscopy methods were also taught at ETH.

He received ample support in his endeavors from his then boss, the chemist Leopold Ružička, who had for some time been convinced of the benefits of physical analysis methods in the field of chemistry. It was Ružička who pushed through Günthard's appointment with the university authorities and representatives from the chemical industry, who at the time funded a significant proportion of the research. Little by little, Günthard brought the whole range of spectroscopy methods to ETH, advancing their development and use in the field of chemistry; this ranged from optical spectroscopy through to infrared and UV spectroscopy. In 1953, the ambitious scientist decided to extend his research into the highly complex area of nuclear magnetic resonance spectroscopy.

As a student at the time, I had little idea of what had gone before me, but the prospect of working in "Günthard's spectroscopy universe" with all these brand-new methods was too much to resist. To start with, however, Günthard gave me a topic that caused more confusion than anything else: "Linear combinations of spin functions associated with irreducible representations". I had actually asked my professor for theoretical work as I was particularly interested in this area. What this task was meant to achieve, however, was a mystery – I found it uninspiring to say the least.

My laboratory savior

When I entered the laboratory on the first day, one of Günthard's employees was stood at a workbench, busy soldering various electronic components together. I was quickly handed a soldering iron and set to work. On numerous occasions I burnt my fingers or gave myself a 300-volt electric shock, but I was gradually learning the secrets of how vacuum tubes could amplify signals and getting to know all about other electronic components used in NMR spectrometers.

The aforementioned employee was Hans Primas, an exceptionally gifted young scientist. It soon became apparent that Primas would be my doctoral supervisor – which with hindsight, was a real stroke of luck. Even today I can hardly name another researcher with such a broadly diversified interest in this area of science, whether in the practical work of a chemical engineer, theoretical physics, or the most labyrinthine concepts in the philosophy of science. Primas would later go on to tackle hugely important issues, but when I saw him for the first time, he was just a man standing at a workbench.

Hans Primas in front of the electronic console of an early NMR spectrometer, in the development of which Richard Ernst was involved. Photo taken around 1958.

At 35, Primas, who came from Zurich, was just five years older than me. He had a different background and did not even have a high-school-leaving certificate. When he was 14, he suffered a bout of typhoid and had to spend several weeks in hospital, followed by a lengthy recuperation in a high-altitude resort in the Swiss Alpine town of Arosa. This resulted in him missing out on a lot of schooling and he had no other choice than to take up an apprenticeship as a laboratory chemist in the analytical laboratory at Oerlikon, an engineering company. He then studied chemical engineering at the School of Engineering in Winterthur, as had Hans Heinrich Günthard before him. Here he excelled, so much so that his teacher recommended him to Günthard, a former student, who was now an outside lecturer at ETH. In 1953, Günthard took him on as a member of his scientific staff, despite the fact that Primas was only able to attend lectures at university as a guest student given his lack of high-school-leaving certificate. Primas went on to enjoy a successful career as a professor of theoretical chemistry and was a leading researcher in the area of quantum theory and a pioneer in the philosophy of science.

At the start of the 1950s, Günthard was quickly impressed by the abilities of his somewhat unconventional employee and asked him to push ahead with brand-new research in the area of nuclear magnetic resonance and build the equipment required. The technically gifted Primas also rose to this challenge, despite – as a chemical engineer – having little experience of electronic engineering. By the time he had taken me under his wing as a PhD student, Primas had already gained some practical experience and together with his employees, had built a first NMR spectrometer from scratch, including all its constituent parts: the probe head, into which the chemical sample to be examined is inserted; strong magnets used to reliably align the nuclear spins; as well as the whole electronic console, with its radio wave transmitters, modulators, pre-amplifier and amplifier, and of course the receiver, which would detect the tiny signals from the nuclear spins. While the relevant equipment was commercially available in the US at the time, it was a sign of Switzerland's scientific ambition that it wanted to develop its own.

I was most impressed by Primas as a person, however. He acted as my lifeline in a lab which would otherwise have thrown me into a potential deep depression. As a student, I was enthralled by Günthard's captivating and clear lectures on spectroscopy. It was he who motivated me in my intense pursuits in the area of physical chemistry and he who was the only one of my teachers to earn my unqualified respect. He was young, forward-thinking and optimistic – all traits I was attracted to. But shortly after beginning my doctoral thesis under the overall supervision of Hans Heinrich Günthard, I found that it was extremely difficult to get close to him or to discuss matters with him.

In his presence, I felt like I was always the young boy struggling to understand what the great master wanted to say. Despite talking in German, half of what he said would be English scientific slang. I remember how once I was stood alongside Primas and Günthard, while the latter was sketching out a theoretical problem on a blackboard. I didn't dare utter a word. It was only Primas, the "insider", who responded, while I was left feeling like I simply didn't belong. I dreaded these discussions and did everything I could to avoid them.

Research group of the Laboratory of Physical Chemistry of the ETH Zurich, 1959: Richard Ernst (2nd row, 3rd from left), Hans Primas (1st row, 4th from left), Hans Heinrich Günthard (1st row, 7th from left), Hans Kummer (2nd row, 8th from left), Fredi Bauder (2nd row, 2nd from right), Peter Bommer (2nd row, 1st from right).

The weekly group meetings with three or four PhD students were somewhat more bearable, even if the professor was in attendance. We sat closely together at a desk, in front of us a sheet of paper on which we would work on a theoretical issue. We would receive a text in advance that would form the basis for our discussions. This at least gave me a chance to prepare something in advance and on occasions, I was even able to make relevant contributions to the discussion. Whether a fledgling thought, a spontaneous idea, even an almost perfectly crafted statement – I would never have dared to express any of these. The focus of these meetings was on mathematical group theory, coordinate transformations and rotations, highly abstract calculations and theoretical observations. While challenging, they trained our brains to adopt a multi-dimensional approach to problem solving, an absolute necessity when it comes to understanding chemistry.

We would often go for lunch in the chemistry bar with Günthard, where we would discuss all sorts of political topics, the leading role played by the US in science and in particular, technology – something that Günthard found particularly impressive. And we believed him. I remember that he never had a good word to say about the situation in the Soviet Union. This was the time when the cold war was brewing between east and west. But when in October 1957 the Soviet Union shocked the world by becoming the first nation to launch a satellite – Sputnik – into orbit and so briefly took the lead in the space race, even Günthard had to acknowledge their military strength. However, this merely served to reinforce his technocratic view of the world: He attributed the Soviets' success to the nation's technological advances and seemingly endless supply of labor.

By the time I reached the end of my doctoral thesis, I had not once had an open discussion with Günthard, whether about personal or scientific matters. He was single-minded and ambitious; his motto was "you can do anything if you want it enough". Günthard was also a very keen sportsman, a trait that ran through his family; his brother Jack Günthard was a famous gymnast in the 1970s and 80s. His PhD students and staff played second fiddle to the goals he had set – he seemed to have little interest in their personal needs and feelings. He was almost dictatorial in setting the subject of my doctoral thesis and even in the 1970s, after I had returned to ETH following my time in the US and was already a professor, he still spoke down to me, calling me "Richi", while I addressed him as "Professor" in return. In his empire, we PhD students were his slaves. Perhaps I am being unfair to Günthard here, perhaps I am projecting the problematic relationship I had with my father in my youth onto him as a person. One thing that is certain, however, is that Hans Heinrich Günthard made a massive contribution to physical chemistry – in particular in the area of spectroscopy – at ETH Zurich, helping lay the foundations for our university's rise to global prominence in this field. But the atmosphere in our day-to-day research work did very little to nurture creativity. By the time I had completed my doctoral thesis, he still knew very little about me and appeared not to care in the least about my ideas, which saddened my greatly. I was happy to have Hans Primas. He was a role model for me, both in a professional and personal capacity. It made no difference that when we were young we would call each other by our last names, a habit that would continue throughout our adult lives, never once switching to being on first-name terms.

I got on extremely well with Primas; you could almost say we were on the same wavelength at the time – this was possibly also because we shared a common passion outside of science: Classical music. We both liked tinkering with things, having grown up with the various engineering and electronic kits available at the time, the "Technikus", the "Elektronisches Experimentierbuch", the "Elektromann", the "Radiomann". There was very little in terms of sophisticated consumer electronics that passed us by and we failed to discuss in detail. One day he managed to get hold of an electrostatic speaker, which having been invented in the 1950s, was the latest craze. It was not long after that I also bought one; they were the best there was, expensive and not that loud, but the sound quality was amazing. Primas sought perfection in everything he did, refusing to compromise on anything.

In the lab, he always had time to crack a joke or discuss things with me, so he clearly must have put in the hard graft at night or over the weekends, as he was extremely creative and quick. His labs became real electronic workshops, littered with measuring instruments, amplifiers, soldering irons, copper wires and all sorts of electronic components. Never far away were the previously mentioned "Red Books" from the Rad Lab, in particular volume 18 about "vacuum tube amplifiers", which we came to revere as our electronic bible.

Apart from me, Primas only had a few other PhD students. One was Rolf Arndt, who was always coming up with light-hearted jokes and wordplays in the lab. Another, Hans Kummer, was later a witness at my wedding, before we lost touch. Fredi Bauder and I went on ski trips and enjoyed hiking in the Graubünden mountains. We would drive or take the train together to various annual science conferences, which at the time were the highlights of the academic calendar. We also liked to place bets on riskier experiments, with a bottle of Campari or whisky at stake. Samples that produced good results were then christened "Campari test tubes".

In September 1959, NMR scientists met at a meeting in Bologna. Back row of the table: Richard Ernst (with glasses), to his right Hans Heinrich Günthard and Rolf Arndt.

Meeting Roswith

If you asked me now about the wild parties and long Zurich nights that other famous scientists experienced – Erwin Schrödinger's amorous adventures and Wolfgang Pauli's escapades in the old town are legendary – I'm afraid I'd have to disappoint you. I was much more the quiet type, something Roswith Tauber recently confirmed to me. She was one of my first girlfriends and we still stay in touch to this day. We almost got married, but we just weren't ready – me much less so than her.

My relationships with the opposite sex were quite complicated and usually also painful too. In some ways, I believe that Freudian psychology got it right when it claimed that a person's sexual desires are one of the most important driving forces in life. Is it not that case that having a treasured partner drives us on to the most creative and highest achievements? And even our constant pursuit of prestige and status in society should improve our chances with the opposite sex, with highly acclaimed scientists no exception here.

I was a late bloomer, however, something that was already clear in high school. Every time I wanted to find a girl to accompany me to the regular dances, I had to engage in a meticulously prepared, almost military operation. By contrast, my schoolmates found it easy to find a partner, or they already had a girlfriend. And once I had found a girl to take with me – usually with the help of my mother or other relatives – that's when the problems really began. On the long walk to where the dance was being held, I would have no idea what I should talk about with my partner. So I started noting down all the possible topics of conversation on a piece of paper before setting off. I would then stick closely to the "plan" – embarrassingly so – and I'm sure you can imagine the stilted conversations we would endure.

Once we arrived at the dance itself, things didn't get any easier. They were often held in restaurants nearby, or they were private balls held at somebody's home, provided there was a big enough living room or cellar. Firstly, we might chat together or have a short poetry recital, but it was then soon down to business. The boys would stand against one wall, the girls against the one opposite and then we would begin to pair off. I was often too timid and would shy away from making a decision. Then came the inevitable – it would be down to the unfortunate last girl standing to endure the hardship of having to dance with me, the most unattractive, but only remaining partner. I was not unathletic – in the military I was a sports officer and in my leisure time I would undertake some fairly challenging mountain hikes – but when it came to dancing, I had two left feet, I was like Bambi on ice. I worked hard to improve, noting down all the steps on (another) piece of paper and during the events, I would subtly make my way to the toilets where I would quickly run through the dance steps I had written down. But when it then came to putting these moves into practice, I would make a right mess out of it, getting both our legs into a tangle and so the evening would frequently end in something approaching a catastrophe.

I hated these dances almost as much as I hated Latin classes. They were both painful to endure and made no sense to me whatsoever! I still remember the days afterwards, when I would become most aware of my shortcomings. It was unbearable. My mental torment would paralyze me and I often suffered from stomach pains and headaches. I had nobody to talk to about my serious problems. I even entertained thoughts about the various ways I could dramatically disappear from this world, but I never really came close to putting any such plans into action.

By contrast, my relationship with Roswith Tauber was actually straightforward, our families having known each other for a long time. I had also spent a lot of time with her cousin Walter Jung at the end of high school, when we were both completing an internship at Hovag in Ems. Following an unhappy romantic relationship, my sister Verena was undergoing psychoanalysis with Roswith's father, the GP and psychiatrist Ignaz Tauber, where she was mainly being helped to deal with her complicated childhood and youth. A very friendly relationship thus developed between our two families, one that still continues to this day.

Through her mother's side of the family, Roswith was related to famed psychoanalyst Carl Gustav Jung. Her mother would occasionally visit Jung, asking him for advice, and later even carrying out research into the field of synchronicity with him. When Jung was already very old, Roswith's parents invited him to Winterthur on several occasions, where he happily answered questions from their wider circle of friends. We were never in attendance, but some of these question and answer sessions were recorded and released to the public, so we later got to hear them. At Roswith's request, Jung later visited the Tauber family to answer questions from the children in private, which saw a trusting relationship develop between the two of them. This culminated in a friendly exchange of letters, in which she asked him questions about life, to which Jung always provided detailed answers.

Roswith Tauber skiing in St. Moritz. At Christmas 1960, the Ernst and Tauber families met on winter vacation in the Engadine.

The Taubers were very different from our family. Roswith's father was young, modern and spent a lot of time with the children. In summer, they would go together to bathe in the river Thur, while in winter they would ice skate together on the Greifensee lake. In the evening, her parents sat together and read the works of Jung and in terms of education, they placed a great deal of emphasis on their children's personal development, believing that they should all have the opportunity to realize their full potential. The atmosphere in the Tauber household was open and communicative, the complete opposite of the traditional practices and air of sincerity to be found in the Ernst family.

Roswith was a pretty young lady with dark blond hair and dark eyes. The relationship I had with her was calm, reserved, platonic. She did not distract me from my work at ETH, or at least I didn't let her distract me. Once when she tried to come and visit me while I was preparing a submission for the Ružička Prize – which was, and indeed still is, a prize awarded at ETH to a young researcher – I was unapologetic in turning her away. Roswith recently told me that she could tell I was extremely goal-oriented even then.

We mainly bonded over our shared passion for classical music. She was a movement coach and had studied since 1960 in Copenhagen under Gerda Alexander at her international school for "Eutony". Eutony can best be described as "balancing tensions" in the body. It aims to develop a person's awareness of their body, so as to have a positive effect on their posture and movement by using the minimum effort to achieve the maximum efficiency. At least that is how Roswith explained it to me. We would write each other letters, in which she told me about the difficulties she was having abroad and I would send her extracts from classical music scores, which went a long way to lifting her mood.

Roswith also influenced me in terms of how I viewed the world and introduced me to Carl Gustav Jung's philosophy and ideas. I even bought a copy of Jung's "Psychology and Alchemy", which I browsed keenly. I never read it in its entirety, in the same way as I have never read any books in their entirety. But I would gladly draw inspiration from the title, summary and certain chapters. The language used by Jung appeared so alien to me, aloof almost. But his thoughts and findings showed me that even the cold science of chemistry had to be understood as a whole, that even chemists are humans with emotions. Jung also showed me that chemistry has a historical and human dimension. His writings on the symbolism of eastern religions and philosophies had a lasting impact on me and most likely paved the way for my later fascination with Buddhist art from Tibet.

When Roswith was on holiday once in Switzerland, we attended a ball at the famed Dolder hotel together. That is when it happened, our first kiss, which I immediately regretted however. Roswith remembers that I had told her that people should only kiss if the situation is very clear for both parties. I was obviously very conscientious when it came to such things, perhaps also too shy. Roswith once wrote to Jung from Copenhagen, asking for advice: "I have a male friend, with whom I share everything I experience – albeit by letter." She asked him whether the story of the fall of man still had any relevance for us, concluding: "I would now like to do everything I can to allow this relationship to flourish." Despite the serious illness he was suffering at the time, the psychoanalyst responded to her letter, explaining that the story of the fall of man still applied because it is a myth, and what is said in myths is timeless and will therefore always be relevant. He then quoted a poem that appeared in Goethe's "Wilhelm Meister": "You lead us into life / You let the wretched man feel guilt, / And then you leave him to his pain / For all guilt avenges itself on earth." He finished by saying that nothing can be crafted without guilt, and that a person can only achieve something if they also pay the costs. Those words made a lasting impression on me.

I later wanted to visit Roswith in Copenhagen, but this wasn't possible. Instead she traveled from the Danish capital to Amsterdam, where I was attending a scientific conference focusing on spectroscopy from 29 May to 3 June 1961. We spent several wonderful days together in Amsterdam, visiting art museums between my presentations to admire Roswith's favorite artist Rembrandt. We then even got engaged. Full of joy, we sent a letter to Jung on our last day together there. "Dear, dear Professor Jung!" wrote Roswith. "We are thinking of your kind reply today, the day of our engagement! ... We send the kindest of greetings from Amsterdam – tomorrow we are both back at work! Yours, Roswith." I added my own message at the bottom: "The warmest of regards and thanks. Yours, Richard Ernst." This could have been the last letter that an already seriously ill Jung ever received, as he passed away just three days later on 6 June 1961 at his home in Küsnacht.

My relationship with Roswith was also drawing to an end, however. The separation was hard and all very old-fashioned. Our relationship had grown like a tender shoot, nourished by the many letters we exchanged as well as the occasional visits – up to the point when we discussed marriage. As was the case with our first kiss, this was not a matter I took lightly. My mother, who after the death of my father had become my main "attachment figure", advised me against marriage. She wrote me a letter, in which she set out her belief that the young Miss Tauber was still too immature and would not make a good housewife. I balked at this advice, but still forwarded the letter to Roswith, which came as a shock to her. The next time she came home on holiday, she came to visit. My mother and I sat opposite her in our living room. I asked her in earnest whether she now wanted to marry me and whether she would be willing to give up her studies in Copenhagen to do so. I had already made known my ulterior motive, insomuch as I intended to take her with me to America after finishing my studies. Given all this – and especially in the presence of my mother – how could she respond other than by turning down my proposal? She later told me that had my mother not been there, she may have made a different choice.

But Roswith also had her doubts. At the time, she had a new group of friends, one of whom – a young lady named Dorothea – had a strong influence over them, telling Roswith that she wasn't yet ready to get married and that America was definitely not the best place for her personal development. So we ended our relationship. Roswith took this badly and was bitterly disappointed, crying for days on end; she would remain unmarried her whole life. Our separation took its toll on both families; Roswith's parents were also sad to see it end. I wrote a letter to the family, attempting to offer some words of consolation. "I am certain," I wrote, "that Roswith is on the right track. Anybody who has such a deep belief in her future cannot fall short of her goals." I get quite emotional when I look back at events, but I accept that things turned out how they did. At the time, our separation was also a form of salvation for me, as I was never so sure of myself and what I wanted to achieve. Nevertheless, Roswith and I remain good friends. Her studies in the field of Eutony went on to form the basis of her future work in art, therapy and education. She developed a method for combining physical movements and Eutony for children, which helped encourage their sense of individuality and their desire to learn. Children were her main area of research. When I look back, this story is like a fading dream, but the battles I had to fight in my life at that time had to be limited to those in the science labs.

Secrets hidden in the noise

While at first, Hans Primas and I spent a lot of time at our workbenches planning and constructing the various key components for new, improved NMR spectrometers, our work also began to drift into more theoretical areas, a necessary step if we were to understand exactly what was happening in our experiments.

In these experiments, you fill a glass tube with a chemical substance and place it between the two poles of the magnet. Such a substance usually consists of molecules in which various atoms are bonded together. Some of these – but not all – have nuclear magnetic spins. In practice, the most important are the hydrogen and carbon nuclei, which align like compass needles along the magnetic field and begin to spin. If a radio wave pulse is now applied to the sample, the relevant nuclei absorb the energy from these electromagnetic waves and then re-release it – but only if the radio waves have the same frequency as the nuclei spinning in the magnetic field. This is the frequency known as the resonance frequency. This is comparable with playing the piano. When a piano string is struck, it begins to vibrate at a given frequency and produces a sound that can be placed on the musical scale. If you wish, you can also then deduce certain properties of the string. With nuclear magnetic resonance, the magnetic nuclei absorb the energy from the radio waves that radiate at this resonance frequency, and re-release them at the same frequency as soon as the radio wave impulse is turned off again. With the right receiver coil, it is possible to record the energy emitted as a signal. The most important thing here, however, is that the resonance frequencies of these types of nuclei, that is to say hydrogen or carbon, undergo tiny changes as a result of the environment of the nuclei in the molecule. These may be only small changes, but it is possible to measure and record them. This then gives us what we chemists call a "spectrum", in our case a nuclear magnetic resonance spectrum: A sequence of different resonance frequencies that tells us a huge amount about the chemical environment of the nuclei. I've always looked at these magnetic nuclei as a sort of "spy", which tells us all the information about the molecules of which they form a part. If we then in turn measure all the resonance frequencies, we can even build up a three-dimensional model of the entire molecule. The method is especially ingenious because this process does not destroy the substance that is being examined, as is the case with mass spectrometry, for example. Neither do we need to use ionizing radiation, as is necessary with X-ray crystallography. All we need are magnetic fields and radio waves.

Workplace of Richard Ernst, around 1958. The laboratory of the NMR researchers was more of an electronics workbench than a chemistry lab; everywhere there were measuring instruments, soldering irons, wires and other equipment.

Using the devices we had built at ETH, we were already able to measure certain spectrums for simple chemical substances. But when our colleagues from the organic chemistry department, mainly

Vladimir Prelog and his young new employee Albert Eschenmoser, would often stop by our lab wanting results that they could actually use, they often left disappointed.

The main problem during these early years of NMR spectroscopy was that the resonance signal itself was often drowned out by the noise from other electromagnetic signals. One reason for this was that the readings from the magnetic nuclei were roughly 1,000 times weaker than other previously known electromagnetic signals; this was ultimately an order of magnitude on an atomic scale. A large part of the disruptive noise was also down to the relatively basic electronics we used at the time – much like an old radio, where it would be hard to make out the newsreader's voice from all the general interference. Human brains are able to filter out the quietest of sounds, easily working out which sounds are important and which not. In physics, for example in the fields of optics or acoustics, there are many mathematical methods for suppressing the noises produced by electronic devices or filtering out the important signals. My doctoral thesis focused on these calculations, which were based on the pioneering work carried out by US mathematician and philosopher Norbert Wiener; our job was to apply them to our problem. We also built measuring instruments and used them to develop experiments which we could use to confirm or discard these calculations.

Our results were interesting from a theoretical perspective, but had very little impact on the practical application of NMR methods. While we did not yet have a computer that could quickly perform the necessary calculations, I did learn how to view the topic of NMR from a system-theory perspective, as Primas showed me how to do. This meant using precise analyses of input and output signals to obtain a wealth of information about what lies between them. In this way, we viewed the noise more as a component of the output signal, which we could pinpoint through careful analysis, rather than an enemy that we had to defeat.

My dissertation armed me with the experimental and mathematical tools I needed to properly understand NMR signals. My doctoral thesis – "nuclear magnetic resonance spectroscopy with stochastic high-frequency fields" – was awarded the silver medal by ETH Zurich and I received a cash prize of 1,000 francs. Even this didn't pass without some embarrassment, however. In the transcript of my thesis I noticed a numerical error in a calculation, albeit a minor sign error, which had no material impact on the theoretical results. But I was flabbergasted by this and it sent a chill down my spine – such a stupid mistake after so many years of hard work! Even today I am still annoyed about this and can't believe it happened. It also goes to show, however, that at the time we were more interested in producing elegant mathematical derivations and pretty formulas than we were in the actual practical implications of our work.

Richard Ernst during his dissertation, here at home in Winterthur. In the background a clarinet, one of the instruments, that the passionate lover of classical music mastered.

At the end, I came to view all my doctoral work as useless and irrelevant. To me, it just seemed to answer questions that had not been asked in the first place. Disappointingly, the results of my work received little coverage in the scientific press, going largely ignored by my fellow researchers. I had had enough of spending my time performing irrelevant mental acrobatics in the lab; I wanted to take my knowledge of NMR and finally do something useful with it. I swore that I would never return to university and applied for a postdoc position in the US. Shortly after, Primas also left the field of NMR and instead focused once more on quantum chemistry. Its principles and consequences, as well as its philosophical importance interested him more than simply applying or optimizing existing technologies. The one thing that really drove him were the absolute truths, a "theory of everything" for the experimental design of spectroscopic methods perhaps, which would explain and then make possible absolutely everything. He had doubts over my decision to stick with NMR. "You can't achieve anything more here. Everything worth doing has already been done," he told me later.

But I stuck by my decision. I was of the opinion that I was yet to achieve all my goals in the field of nuclear magnetic resonance, so I busied myself preparing for my time in the US. But first of all, I wanted to find a wife, with whom I could start a family.

Silver Linings over the Pacific, 1963–1968

Choosing a partner

After finishing my doctoral thesis, I had the feeling that I was a performer without an audience when it came to my scientific work. Both Hans Primas and I had reached the limits of our work and we were still ultimately able to present a coherent theoretical concept for our experiments. The only problem was that none of this seemed to make any actual impact. We weren't able to provide chemists with a way to produce better NMR spectrums and quantum physicists had no interest in our work. Primus and his staff at ETH had built their own NMR spectrometer that could be used to carry out simple chemical analyses, something I had helped optimize. At the time, the devices were even produced commercially by Zurich electrical engineering company Trüb, Täuber & Co. AG and sold to various research institutes and chemical companies throughout Europe. However, the devices simply weren't able to make any use of my theoretical work. What's more, the Californian company Varian Associates had since brought a product to the global market – the legendary A-60 spectrometer – which had helped establish the use of NMR as one of the most important chemical analysis processes in chemistry labs around the world.

On top of that, I found the atmosphere at ETH somewhat depressing. I was being asked to produce results according to a strict timetable; the structure was rigidly hierarchical and it was an unwritten rule that any of our own goals came a distant second to the goals of our department head. This was the complete opposite of the environment in which I believed creative research could flourish. All of this meant that creativity stemming from my unbridled passion for discovering new things – that almost childlike joy – was simply impossible. Placing anything off limits, in particular free thought, and the imposition of strict rules will always stifle creativity. Anybody who sticks to such rules cannot be a researcher, and anybody who doesn't push those limits, even less so. If you are not allowed to question the obvious and have to accept everything as it is, then you cannot be creative – it's as simple as that. At the time I missed all of these things at ETH. We were stubbornly plowing onwards, never looking left or right. The fact that all of this meant I was building up an outstanding fundamental knowledge of my field – which is one of the basic prerequisites for creative research – was something I failed to fully appreciate at the time.

The path to the US for ambitious researchers in scientific specialty fields was already well trodden; even a good career in industry had by then already become difficult without a postdoc achieved abroad. I felt the need to break out of my ivory tower and so wrote only to companies, sending off around twenty letters of application in all. I courted well-known companies such as Hewlett-Packard, the then electronics giant Radio Corporation of America (RCA), and Bell Labs, but also oil companies in Texas, which were of course very interested in the analysis of organic chemical substances. Some of these companies invited me for interviews, even offering to pay my airfare! So I planned a trip to the US for February 1963 to visit these firms. The Californian company Varian Associates, which would go on to employ me as part of my postdoc program and where I would make my scientific breakthrough, was not actually one of the companies on my list.

At this moment – it must have been some time around the end of 1962 – I received a written invitation from Arnold "Noldi" Renold to attend a musical evening in his home. At the time – long before night clubs became popular – these evenings would attract young people from the circles in which we moved and were particularly popular with teachers. We would meet mainly to chat, but there was also plenty of classical singing and music. Noldi Renold was a young teacher with a wide range of interests; he had a wide social network in Winterthur, was extremely musical, and was a member of a choir. I was pleased to be in attendance, but also anxious, as was always the case at these events. Secretly, I was thinking about finding a companion, someone who would accompany me to the US and cook my meals for me in this strange land. I was scared of having to go alone and risk being surprised by the aggressive American women we had heard so much about, women – who in my mind – were the polar opposite of me. It goes without saying that this was a childish stereotype, but I'm sure my fear of unexpected social encounters also played a role in this. So the idea of finding a nice Swiss girl from my home town seemed to be a calculated risk that was well worth taking.

The plan was for me to play a piece on my cello at the musical evening. I tuned my instrument and took the number 3 bus to Bettenstrasse 153 in Winterthur-Veltheim, where Noldi Renold lived in a semi-detached house, as was usual for teachers at the time. There were around ten people invited to the evening; you might even term it a "party" nowadays. We sang together and I played my cello. I can't remember who else had brought an instrument with them, but I can remember the journey home. I again took the bus, but this time I was not on my own: A young primary school teacher named Magdalena Kielholz, who Noldi Renold knew from school or his choir, got on the same bus. It was the first time that we talked together, I assume about music and singing. When Magdalena got off two stops before mine, I also got off with her, before walking the rest of the way home with my cello.

When we met in 1962, the main impression she had of me was being "dochtig", a Swiss-German word used to describe someone, fairly or not, as being somewhat ungainly. She recalls now that I was very reserved, shy, clumsy, and almost a little depressed. Our conversations on the telephone were hard work. At the time, speaking over the phone was a key part of any relationship as it was unusual to meet up all the time before being married. Because spontaneous conversation was something I still struggled with, I would note down on a piece of paper what I wanted to say to Magdalena before calling her. And I stuck rigidly to my notes, being sure to say everything I had written down. By contrast, Magdalena had no difficulty engaging in spontaneous conversation. When it all got too much for me, I would hold the receiver away from me at arm's length and wait a moment before starting the conversation again. I would carry on reading my notes until I got to the point where I had no idea what to say.

Magdalena remembers how day after day, I would take the train to Zurich, climb the Weinberg footpath and hide away in my lab, grappling with my unworldly problems. She says I would often glance at the pretty nurses coming and going from the hospital over the road, reminding myself how I was still yet to find the wife I so desperately wanted. My relationship with Roswith Tauber had ended several months previously. I was shy, so shy in fact that I even turned down a research assistant's post at ETH simply because I didn't want to suffer the ignominy of having to speak in front of students. The idea of giving a lecture petrified me. Anybody asking me about my work would simply be told that only a handful of people in the entire world would understand it, that it's anyway just a game, the purpose of which I didn't understand even myself, that I'm nothing more than a useless academic.

I had become a real loner: Almost thirty years old and still living with my mother. She cooked my meals and washed my clothes, while I took care of the various DIY tasks that needed doing in our large house on Gottfried-Keller-Strasse. My rebellious sister Lisabet had since moved out, leaving as quickly as she could. She had barely finished high school when she traveled to Florence to continue her education. Lisabet once told me that she did not at all approve of the way I had taken over from where my father had left off, saying that I had become a little too "domineering" at this time. Verena, my closest companion as a child, had become a nurse and was away working somewhere in Scandinavia on a work placement. She had already met her future husband and their wedding was already planned. My mother also wanted for me to get married, but whenever I met anyone who I considered a serious "candidate", she would not approve. Any potential daughter-in-law had to fulfill all sorts of requirements; my first girlfriend, Roswith Tauber, did not meet her high expectations.

I was very fond of Magdalena, but I had to proceed with caution. At first I would send her short messages by post. At Christmas, I sent her a parchment with the start of the St. John Passion in Latin; Magdalena loved it! In January or February, I set off to the US for a round of interviews with potential future employers. The trip in itself was a success, resulting in several job offers, but I couldn't get Magdalena out of my mind while I was away. At one stop on my tour I visited an art museum and sent her a postcard featuring a self-portrait by Rembrandt. I probably thought that anyone who loved music would also love Rembrandt; after all, this was the case with Roswith. It turned out that my assumption was not too wide of the mark, as Magdalena found it charming. She thought that a man who didn't just have "sport on the brain" all the time and instead had a taste for music and art was extraordinary and attractive enough as it was.

Magdalena is three years younger than me and comes from a family of teachers in Zurich. Her parents were teachers, her brother and three sisters all trained in the profession, and she herself was a teacher. Her father Paul Kielholz was a well-known figure, in teaching circles at least. He had a real influence on my wife – as he did on all his children – not only in terms of the genes he passed on to her, but also in his ideas and aims in life. I would like to talk a little about him as he also indirectly, that is to say via my wife, had an influence on the spiritual and idealistic atmosphere in our family.

Paul Kielholz was born into a poor family in Zurich in 1908. His father was a bookbinder, but lost an arm in an accident at work, meaning that he was later only able to work as a telegram deliveryman. Paul Kielholz first trained as a draftsman and worked in an architectural practice. After going traveling, which was common at the time, he attended the teacher training college in Küsnacht, where he met Magdalena's mother Ida Meierhofer, who he then went on to marry. She was a farmer's daughter from Weiach in the Zürcher Unterland, the oldest of seven siblings. She was the only child to be given the chance to receive a better education, for which her family had to pay boarding and school fees. The plan then was for her to earn money as a teacher, which would then pay for her younger siblings to learn a "proper" trade.

At the teacher training college in Küsnacht, Paul Kielholz quickly became an opinion leader in his class, mainly because he was older than all of his colleagues. He was extremely musical and organized a music group in the lunch break, playing the guitar and singing songs that he had learned while traveling. Following a temporary post at a secondary school in the countryside, he took a job in the Limmattal school district, where he became a teacher in 1937, remaining there until his retirement in 1975. From the very start, he began to make use of child-focused teaching methods, making the lessons as true to life as possible. He taught botany by relating the lesson to the weeds in the playground, excursions to local places of interest were a regular feature of his history lessons, and his maths lessons were full of everyday examples. Many of these methods, which originated in European progressive education, were initially tested in his schoolhouse; following many years of development, they were then ultimately recommended and implemented at teacher training colleges. This gave the methods a "Swiss" touch and meant they were better received in schools. Teachers in the working-class area of Zurich were opposed to the excessively logic-based approaches of the time as well as the cold, high-register language used in text books. Progressive teachers instead were campaigning for a more vivid way of teaching, which children could relate to their everyday lives. Paul Kielholz soon came to be seen as a pacifist and communist among traditional teachers, but this did not discourage him.

Magdalena was the oldest of five siblings. Because her mother returned to work soon after giving birth to her youngest sister, which at the time was highly unusual, Magdalena had to look after her younger siblings – a tradition going back to her mother's roots as a farm girl, when this was commonplace. Magdalena did not find this an easy responsibility to bear. While her parents had employed an Italian migrant worker as a nanny, their oldest daughter still had to shoulder a heavy burden, not least because she was supposed to be preparing for her high-school-leaving exams – she was aiming to pass the challenging version, which included ancient Greek. She therefore had a difficult relationship with her mother, who was on the one hand, very progressive – nowadays she would be considered a feminist. When Magdalena's brother was born, however, it quickly became clear that he was her favorite child. Magdalena's main way of escaping this burden was her choir, a hobby that she pursued with great passion and commitment. She still now has a glint in her eye when she remembers back to singing in her family, sometimes with her brothers and sisters.

When we met, Magdalena was already living and working in Winterthur. She had found a job at the Schachen-Schulhaus in the north of the town. She lived on her own for five years in Winterthur, although she originally would have preferred to stay in Zurich. But she was happy there; she joined a new choir and studied the violin under renowned musician Aida Stucki, often helping out in the choir for vocational students at the Winterthur conservatoire. This is where she got to know Noldi Renold, who invited her to the musical evening where we first met. He later revealed an ulterior motive to inviting us both: He wanted to introduce us as he knew about my failed relationship with Roswith and knew that Magdalena, too, had previously been unlucky in love.

Not long after our first meeting, I revealed to Magdalena that I would soon be emigrating to America and would very much like her to accompany me. She was somewhat shocked and so sought her mother's advice. Her mother, however, encouraged her, telling her to "go to America, seize this opportunity!"

Richard Ernst and Magdalena Kielholz at their engagement at Whitsun 1963 in Konstanz/D.

When I returned from my round of interviews in the US and would not stop talking about the "land of the free", Magdalena was still not convinced, however. She always dreamed of revisiting Israel, having previously toured the newly founded country with her choir, where they visited various kibbutzim to perform concerts. Magdalena was fascinated by the country and its people – in Switzerland, there was widespread positive sentiment towards Israel at the time. She was more skeptical towards America, dreading the fast-moving, superficial way of living that she thought would await her there.

Magdalena and I married at the church in Oberwinterthur on 5 October 1963. Her choir serenaded us and the reception was held at the Drachenburg hotel in the village of Gottlieben on the shores of Lake Constance, which at the time was a popular destination in eastern Switzerland. It goes without saying that my mother had thoroughly scrutinized my future wife beforehand. Magdalena remembers how on one occasion, my mother's father, grandfather Brunner, who was also a teacher, arrived unannounced in her classroom, wanting to check how she dealt with her pupils, what sort of lessons she gave – in short, whether she was suitable. I cannot remember the exact outcome of my mother's "assessment", but it obviously didn't stop me proposing to Magdalena. Nevertheless I told her from the start that I would never have a great deal of time for my family. She accepted this as our marriage suddenly seemed to be a good way out of her difficult family life.

I talk about this in so much depth here as I am aware of just how much our bond impacted my scientific work. Magdalena gave me the space I needed, she understood my frequently inflexible work ethic, she did everything that needed doing at home and with the children without complaining. Looking back, it seems that our courtship, marriage and move to the US all went so quickly that we had no time to fall in love. It was as if we had simply skipped the phase of infatuation and had entered day-to-day life before we knew it. My wife always describes us as two stones in a wild mountain stream, encountering each other by chance and then simply remaining together. This bond between us has now lasted more than fifty years. Magdalena was happy to have escaped a difficult family situation and was keen to experience something new, while I was pleased to have found a wife with whom I could start a family. And the fact alone that I had finally got married and had a wife allowed me to focus on what was important in my life, something I can't thank Magdalena enough for. "I can finally work how I want to," I told her on numerous occasions in California.

On October 5, 1963, Richard Ernst and Magdalena Kielholz were married in the church of Oberwinterthur: Ida Kielholz, Magdalena and Richard Ernst-Kielholz, Irma Ernst, Paul Kielholz (from left to right).

Arriving in the "land of plastic smiles"

Before we got married, Magdalena made me promise her one thing: That we would travel to American by ship. This was the only way that her soul would have time to adjust to the move, she told me, not least because we were traveling to a country which she believed would be far more hectic than Switzerland, more superficial: She described America as the "land of plastic smiles". So a few days after our wedding, we took the train to Rotterdam, where we boarded the SS Rotterdam passenger ship. This was our honeymoon, but because it had been booked at short notice, we could only get a cabin where you could not escape the noise of the ship's engines. The crossing took seven days. One day, there was such a violent storm that the captain had to shut down the main engines, just to keep the ship on course. While this had the benefit of us being able to enjoy some peace and quiet in our cabin, I suffered from bad seasickness. It was on this day, my wife remembers, that we probably enjoyed our last dance for some time, a cha-cha-cha. I can imagine I don't remember this as I was seasick at the time. It came as a real relief when the ship docked in New York and I could set foot on dry land once more. We then flew across the country by plane, where we landed one sunny afternoon in San Francisco.

Arrival in the country of opportunity: Magdalena Ernst on the bridge of the "SS Rotterdam" as she enters New York in November 1963.

After landing, I called Weston Anderson, my future boss at Varian Associates. He had offered to pick us up from the airport, but I told him this was unnecessary as we had a map of the city and I wanted to be truly independent and make my own way in this new world. On the same day, we bought a used car from a shady car dealer on the then famous Van Ness Avenue, a Ford Falcon for 3,700 dollars. I felt a bit uneasy doing so as I had only passed my driving test a few weeks previously – on my second attempt. My Swiss driving instructor told me that I didn't have a talent for "technical things", and this just one day after I had been awarded the silver medal for my dissertation, as part of which I had mastered a technology that was most likely decades ahead of its time and which my driving instructor would not have known even existed. And at home, I was always the one who took care of all the manual repair work. It's no surprise that my driving instructor's offhand remark wounded my pride somewhat.

Magdalena Ernst on a trip in the countryside of California. Richard Ernst had bought the Ford Falcon immediately after arriving in San Francisco. The car proved to be very susceptible to repair.

The drive from San Francisco to Palo Alto in our newly acquired Ford Falcon was a real adventure, mainly due to the miserable condition of our used car. It was no time at all before one of the tires was flat, not long after that the battery died, and just a week later, our engine started leaking oil everywhere. We had barely arrived in Palo Alto and we had to get the car repaired in a garage. When we got the bill, we quickly realized that the car dealer in San Francisco had taken us for a ride. We had no idea that the insurance premium we had paid only covered replacement parts!

Back then, Palo Alto was a small American suburb, spread out along the palm tree-lined Camino Real, the historic route that

Spanish missionaries would take along the Pacific coast. Even at the time, the road was a busy highway, with one gas station after the other. The town is a little more than 50 kilometers south of San Francisco and is home to the world-famous Stanford University, on the doorstep of which was located Varian Associates, the destination of my long journey. Nowadays, this is in the middle of Silicon Valley, the epicenter of the computer revolution – a revolution that few knew about at the time. We received a warm welcome from Weston Anderson – known by everyone simply as Wes – and his wife Jeannette. Wes is an amazing person: Independent, free, kind, and most of all, an outstanding, creative scientist. We all become friends from day one, and have remained so ever since. I remember one episode at an airport: I had taught Wes how to play the recorder and once, when we were waiting for a flight, he took out his recorder and just started playing, not caring about all the people who heard him and turned to stare. We would often go hiking together in the Sierra Nevada or in one of the stunning national parks California had to offer – I enjoyed all of our excursions together.

The Camino Real was the lifeline of the young Silicon Valley of the 1960s. The road runs along the historic routes of the Spanish missionaries through California.

Weston Anderson, head of the science department at Varian Associates and chief of Richard Ernst. Photo taken around 1969.

The nucleus of Silicon Valley

The fact I landed a job at Varian Associates is one of those strokes of luck in a researcher's life that you have no influence over. I had written to the company, but was not invited to visit them at their headquarters in Palo Alto when I was in the US in early 1963. It was only later that a representative of Varian contacted me in Zurich, where the company ran a so-called "application laboratory" at ETH. Said representative, Warren Proctor, was himself a scientist and wanted to find out whether I might be interested in a postdoc program at the company. He invited Magdalena and me to dinner at the Rüden guild house to discuss the details – a prestigious address if ever there were. It was in June 1963 that Hans Primas wrote me a wonderful reference that still flatters me to this day. We had worked closely together at ETH for seven years and valued each other's input greatly. "During that period of close contact," wrote Primas, "I learned to appreciate his sincere character, his integrity and his outstanding scientific capabilities." He continued: "His research work is characterized by an excellent combination of high experimental skills and profound theoretical insight." Primas even went on to praise my teamwork, a skill that even then was clearly sought after. While I was definitely still somewhat of a loner in my private life, when it came to research I was always a team player provided the right conditions were in place. During my dissertation, I put together a team with Hans Primas; later on in California it was mainly Wes Anderson; and then at ETH Zurich, my own research group.

Hans Primas' letter of recommendation and the impression I had made on Warren Procter paid off: Varian Associates wanted me to work for them and they offered me a job in their Instrument division, which was the department responsible for developing NMR equipment. Varian were promising me almost exactly what I was after. I could continue the work that I had started in Zurich with the only difference being that Varian had commercial objectives. This gave me added motivation as following my stay in my ivory tower at ETH, I finally wanted to do something that seemed to benefit society as well.

Varian was essentially a text-book example of a company harnessing the latest high-tech science in the first half of the twentieth century. Together with the development of computers, the NMR spectrometers successfully developed by Varian were placing chemistry on a new footing. The company is also a perfect example of how the best researchers from universities and industry could work together in the post-war years to produce mutually beneficial results. It is worth taking a look back at the history of this unique company.

The name Varian comes from Russell Varian and his younger brother Sigurd, the sons of Irish immigrants. Russell was the inventor, Sigurd the pilot. Russell Varian studied in the physics department of Stanford University in the 1920s, but his academic career stalled after this. Because he had reading difficulties, did not speak any German or French, and because he was not an outstanding mathematician, he was not admitted to the university's doctoral program. Instead, he remained a research assistant at the institute. His brother Sigurd had initially trained as an electrician and went on to become a pilot at Pan Am. As such, he was well aware of the as yet unresolved problems when it came to flying safely in poor, low-visibility weather conditions. Russell identified that radio waves, which could pass through clouds and also help pilots fly at night, could be a potential solution to Sigurd's problems. Together with Russell's former classmate Bill Hanson, who was now a professor of physics at Stanford, in 1937 the Varian brothers invented the "klystron", a linear-beam vacuum tube which could amplify electromagnetic signals and be used in radar applications, for example. At first, there was limited interest in their findings in the US. However, word quickly spread to England, where scientists had already been hard at work carrying out research into radar technology; this innovation all the way from California was a success. The klystrons were ultimately incorporated into the on-board radar systems installed by the British in their fighter aircraft, which played no small part in helping their pilots overcome the Nazis in the Battle of Britain. They were subsequently developed further for use in particle accelerators, telecommunications, and microwave technology.

The klystrons earned the Varian brothers and the university, which was contractually entitled to a percentage of profits, their first income. It was not until 1948, however, that Russell and Sigurd – together with some like-minded scientists from the so-called microwave research group at Stanford University – founded their own company: Varian Associates. During the war, researchers had been serving their nation and were spread across the entire country. When they then returned, positions at universities in their field were suddenly a rarity: Their skills appeared to no longer be relevant during peacetime. But shrewd researchers and business-minded engineers quickly realized that the knowledge they had acquired carrying out research during the war could in future be applied to benefit the general public and to help society progress as a whole. "Swords to ploughshares" was the prevailing theme, no less so in microwave technology than in nuclear research.

In essence, Varian became a home for scientists and engineers from Stanford University's physics department who could no longer make as much of a living working in public-sector research. Their relationship with academic researchers remained very close, however. Thanks to the work carried out together with Swiss-born scientist Felix Bloch, who had been researching and lecturing at the university since 1934, the company specialized in NMR spectrometers. Bloch had discovered nuclear magnetic resonance in 1946 and even then, had identified the potential it could have in chemical analysis. He went on to receive the 1952 Nobel Prize in Chemistry for his work in this area. However, he was more interested in fundamental research; for him, nuclear magnetic induction, as the phenomenon was known at the time, was initially just another step on the way to uncovering more about nuclear physics. But Russell Varian urged Bloch to patent his discovery and to be sure that his patent application mentioned the uses the technology could have in chemical analysis. The patent became an important source of income for the company, even though a practical application for the technology was still some way off in the early years. It's therefore fair to say that Varian Associates was one of the earliest examples of a university spin-off.

In 1953, the company was the first to relocate to the Stanford Industrial Park, which had recently been established by Stanford University and is considered the nucleus of Silicon Valley. It's now known as the Stanford Research Park and is home to 150 research-based companies, which employ 23,000 people. The Varian brothers knew even then that a company's most important resources were its inventive engineers and highly specialized scientists. The two socially engaged entrepreneurs believed that it was important to foster a corporate culture in which creativity could thrive. Their business was more like a research lab, located coincidentally in front of – rather than behind – the university's gates. Making a fast buck was not a priority and the issue of controlling costs was rarely discussed. The aim was for the engineers and scientists to be as free as possible to research their own ideas as they saw fit. It also went without saying that wherever possible, their work should ultimately result in patents that could be monetized or other commercial applications. Thanks to support from government – and still also military – research grants, the company grew quickly and ten years after its creation, already employed 200 scientists and engineers.

Unfortunately, both of the Varian brothers died shortly before I arrived in California. Russell, the inventor, suffered a heart attack in 1959 while hiking in Alaska and died aged 61. Sigurd, the pilot, crashed his private plane into the Pacific while on a night flight aged 60. Despite their premature deaths, the relaxed and research-focused atmosphere they had helped create was still very much in evidence when I arrived in California in 1963. From day one, everyone was on first-name terms, no matter whether you were a high-ranking manager or junior scientist. Casual conversations and productive coffee breaks were part and parcel of everyday working life – contrary to what some people believe, this wasn't invented at Google! As a newcomer, who had only recently escaped the formal, hierarchical practices of traditional European universities, I was amazed and inspired by this refreshing approach to research in the New World. Even in the late 1960s, we scientists and engineers still formed the largest, most important group of employees at Varian. Subsequently, the founding generation with its scientific background was gradually overtaken by business-focused managers and lawyers, whose main priority was to generate the necessary cash flows. This was to play a key role in how my work would later be exploited.

First breakthrough thanks to the Fourier transform

Wes Anderson had joined Varian a few years before I crossed the Atlantic. He too had written his doctoral thesis on NMR, in fact in the lab run by Felix Bloch. In 1954, he traveled with the recent Nobel Prize recipient to Geneva, where Bloch had been appointed the first Director-General of CERN, the newly founded European Organization for Nuclear Research. After just one year, Bloch had already had enough of the myriad of administrative tasks landing on his desk and returned to his research team in California. Wes Anderson soon followed him back to Stanford – at Bloch's urgent recommendation: He had to be surrounded by his best scientists, both for his and their benefit, he wrote to a manager at Varian, for whom Bloch had since become a consultant. So in autumn 1955, Wes Anderson was hired as a researcher at Varian. His job was to optimize the existing NMR spectrometers, often in cooperation with scientists working in chemistry labs around the world, who were becoming ever more interested in this new method.

One meeting that made a lasting impression on me occurred at the start of my postdoc in California with the great Felix Bloch. A few days after arriving, I was due to give a presentation at an internal company seminar. These events were a weekly occurrence and were held so employees could exchange ideas and learn more about others' work. Researchers had to deliver a presentation about their work or a study that was relevant to them; it was common practice to invite guests from other companies or universities.

So just days after landing in the US, I was already being thrown in at the deep end. I remember working on my speech right until the last minute; even on the flight from New York to San Francisco, I was sat in my seat with notes piled high on my tray table. Nevertheless, I was anxious and scared given how I had trouble speaking in front of an audience. When on that Wednesday I approached the lectern in the seminar room, I noticed that Felix Bloch himself was in attendance. To be honest this came as a shock, as I was in awe of the man.

Bloch was a prominent figure in the area of NMR, as indeed he was in quantum mechanics in general. Born in Zurich in 1905, he studied physics at ETH Zurich in the 1920s, when the university was an epicenter of the quantum mechanic revolution. He attended lectures given by famed Dutch physicist Peter Debye; he was present at conferences held by Erwin Schrödinger, where he explained his newly developed wave equations. In 1928, Bloch moved to Leipzig University, where he earned his doctorate under Werner Heisenberg, the father of the uncertainty principle, and where he later became an outside lecturer. In between, he returned for around a year to Zurich, where he worked together with Wolfgang Pauli, and his CV also featured fellowships with Niels Bohr and Enrico Fermi. But after Hitler seized power in 1933, the situation rapidly deteriorated for many young Jewish physicists in Germany: Bloch's name appeared on the "List of Displaced German Scholars", despite his being a Swiss citizen. This meant he wanted – or indeed needed – to leave Germany in 1934 and so emigrated to the US, where he found a home at Stanford University, even though he only knew the university from hearsay. This is where he took up his first chair in theoretical physics. Before the war he conducted important experiments in nuclear physics, but following the outbreak of conflict – after having taken US citizenship – he served his adopted nation by working on the development of nuclear weapons as part of the Manhattan Project, albeit somewhat unwillingly and therefore only for around twelve months. He subsequently worked in the Rad Lab at MIT on research into radar technology. Once back in Stanford, his experiments resulted in the discovery of NMR in 1946. This led to him becoming a scientific consultant at Varian, which meant he often took part in the weekly seminars at the company.

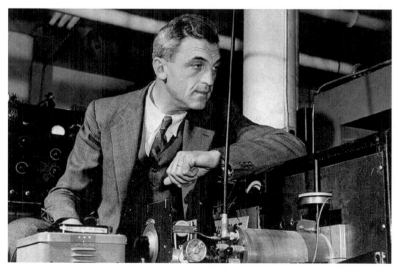

The Swiss physicist and Nobel Prize winner Felix Bloch in the 1950s. Bloch was a consultant to Varian Associates. Richard Ernst often met Bloch at staff seminars.

In my first seminar at Varian, with Felix Bloch in the audience, I was as nervous as a schoolboy. It was a rather dry presentation on my part – I spoke on the topic of operators, super-operators and eigen-operators, which are mathematical constructs I used in my dissertation to analyze NMR signals. After I had finished speaking, Felix Bloch spoke up: "What you managed to do in Zurich," he commented approvingly, "was something really special." I was so proud when I heard his comments. After the seminar he came up to me and we moved to the library to continue our discussion. We searched out a couple of scientific papers on the topic and examined a number of technical questions. It was certainly a Wednesday to remember! I now really felt that I belonged in my new job.

Working at Varian was intense and challenging; during the day we would perform experiments in the labs, while in the evening I sat at home immersing myself in theoretical principles. I worked primarily with my boss Wes Anderson, who helped me greatly. Our pet project was to optimize the existing NMR measuring instruments. At the time, the process was still extremely slow. You may remember that the nuclei which are aligned and spinning in a magnetic field are exposed to radio waves until they absorb a certain frequency. The resonance frequencies had to be gradually adjusted step by step, just like if you were adjusting a radio dial, searching for one station after the other. Recording an NMR spectrum was a case of slowly turning the dial and trying to capture all the interesting phenomena. Wes Anderson and I decided to try a different approach, however. We exposed the sample to a broadband pulse and then simply observed what happened. This principle can best be explained using the example of a piano. If you play a scale, you press one key after another and hear one note after another. In theory, however, you could press all the keys at once; you would then hear every note being played at the same time. While this might sound dreadful, the noise produced would contain all the notes you wanted. This is clearly a pointless experiment on the piano, but if you wish to test a complex chemical compound, this method can save huge amounts of time. If you expose a sample to every frequency at once, all magnetic nuclei also react at the same time.

While the measuring instrument will also therefore register a variety of resonance frequencies, which at first glance might look like background noise, you can use a "filter", analyze the signals and separate out those that are important from those that can be ignored. This filter – or more precisely, algorithm – is a mathematical equation called the Fourier transform. It harks back to the French mathematician Joseph Fourier, who at the start of the 19th century, was studying the way in which heat is conducted in solid objects. In doing so, he developed this mathematical operation, which is often also used in the fields of optics or acoustics. It can also be used to derive useful information from a large number of signals or data streams that are otherwise difficult to interpret. In principle, we were therefore separating the process for measuring the signals from the process for analyzing the signals. Performing such an experiment now takes a matter of milliseconds, while previously we would have had to spend hours working through the different frequencies. By applying the Fourier transform, Wes and I managed to take a muddled series of measurements and produce a simple spectrum, which provided clear, chemical information about the substance being tested.

This theory was all well and good, but I needed a way to put it into practice: For this I needed a more powerful amplifier. When I asked a colleague from the lab next door whether he could build me one, he seemed confused by my request: Of course he could, he said, but he couldn't work out what this had to do with NMR. The old method needed a radio wave generator with just a few milliwatts of power, because we only wanted to slowly test the entire frequency spectrum step by step. I had something else in mind. I wanted to fire a short, extremely powerful broadband pulse at the sample to excite all the nuclear spins, and for this purpose, I needed an amplifier with at least fifty watts of power.

So my colleagues set about building me a new amplifier, which I then connected to my NMR spectrometer. The next several weeks were spent hiding away in my lab conducting one experiment after the other.

Analyzing the signals was proving to be a tedious and time-consuming affair. First of all, you had to transfer the signals, which had been spat out onto seemingly endless pieces of paper, into punched cards and then onto magnetic tapes. Only then could they be read by one of the early IBM computers, which would perform the Fourier transform and then finally print out the resulting spectrum on a rather primitive printer. The whole process took days, mainly because the only computer in the company was reserved for the accounting department. It was only after processing all the staff salaries and completing the inventory that we were allowed to use it to analyze our NMR data.

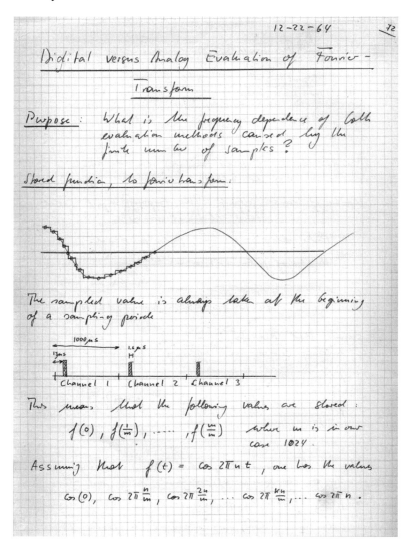

Digital versus Analog Evaluation of Fourier-
Transform

Purpose: What is the frequency dependence of both evaluation methods caused by the finite number of samples?

Stored function, to fourier transform:

The sampled value is always taken at the beginning of a sampling period

Channel 1   Channel 2   Channel 3

This means that the following values are stored:

$$f(0), f(\tfrac{1}{m}), \cdots , f(\tfrac{m}{m}) \quad \text{where } m \text{ is in our case } 1024.$$

Assuming that $f(t) = \cos 2\pi u t$, one has the values

$$\cos(0), \cos 2\pi \tfrac{u}{m}, \cos 2\pi \tfrac{2u}{m}, \cdots \cos 2\pi \tfrac{ku}{m}, \cdots \cos 2\pi n.$$

79

1. <u>Digital evaluation</u>:

$$\boxed{\mathcal{F}(n) = \sum_{k=0}^{m-1} \cos^2 2\pi \frac{kn}{m} = \frac{m}{2} + \frac{1}{2}\sum_{k=0}^{m-1} \cos 4\pi \frac{kn}{m} = \frac{m}{2}}$$

This means the result is independent of n.

2. <u>Analog evaluation</u>

$$\mathcal{F}(n) = \int_0^1 g(t) \cdot \cos 2\pi n t \, dt$$

$$= \sum_{k=0}^{m-1} \cos 2\pi \frac{kn}{m} \int_{\frac{k}{m}}^{\frac{k+1}{m}} \cos 2\pi n t \, dt$$

$$= \sum_{k=0}^{m-1} \cos 2\pi \frac{kn}{m} \frac{1}{2\pi n}\left\{ \sin 2\pi n \frac{k+1}{m} - \sin 2\pi n \frac{k}{m} \right\}$$

$$= \sum_{k=0}^{m-1} \cos 2\pi \frac{kn}{m} \cdot \frac{1}{2\pi n} 2 \cos 2\pi n \frac{2k+1}{2m} \sin 2\pi n \frac{1}{2m}$$

$$= \frac{2}{2\pi n} \sin 2\pi n \frac{1}{2m} \sum_{k=0}^{m-1} \cos 2\pi \frac{kn}{m} \cos 2\pi n \frac{2k+1}{2m}$$

This gives a complicated result with a further unwanted attenuation at high frequencies. It is better to phase-shift the computing cos by 1/2 channel length.

$$\mathcal{F}(n) = \int_0^1 g(t) \cos 2\pi n \left(t - \frac{1}{2m}\right) dt$$

$$= \sum_{k=0}^{m-1} \cos 2\pi \frac{kn}{m} \int_{\frac{k}{m}}^{\frac{k+1}{m}} \cos 2\pi n \left(t - \frac{1}{2m}\right) dt$$

$$= \sum_{k=0}^{m-1} \cos 2\pi \frac{kn}{m} \frac{1}{2\pi n} 2\left\{ \cos 2\pi n \frac{k}{m} \sin 2\pi n \frac{1}{2m} \right\}$$

$$\boxed{\mathcal{F}(n) = \frac{2}{2\pi n} \sin 2\pi n \frac{1}{2m} \sum_{k=0}^{m-1} \cos^2 2\pi \frac{kn}{m} = \frac{1}{2} \frac{\sin \pi \frac{n}{m}}{\pi \frac{n}{m}}}$$

In this case, the result depends slightly on n.
The extreme case is $n = \frac{m}{2}$. Here we have

$$\mathcal{F}\left(\frac{m}{2}\right) = \sin{}^{\pi}\!/\!_2 / \pi \quad = 0.3183$$

$$\mathcal{F}(0) = \qquad\qquad = 0.5000$$

If this phase shift is not made $\mathcal{F}\left(\frac{m}{2}\right)$ goes to zero.

Practically the phase shift could be made in the following manner:

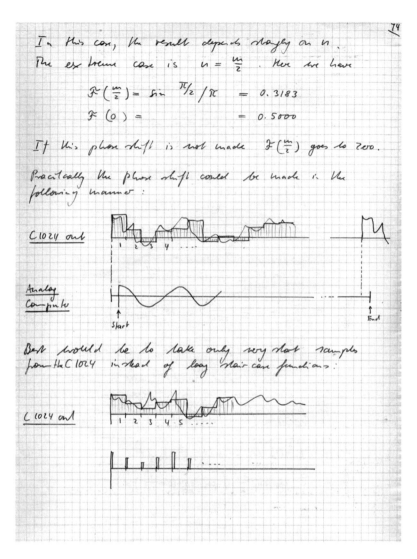

C 1024 out

Analog
Computer

Start

End

Best would be to take only very short samples from the C 1024 instead of long stair case functions:

C 1024 out

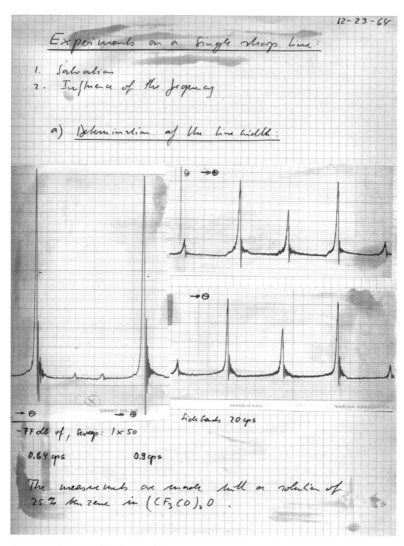

Excerpts from Richard Ernst's laboratory journal from 1964. In the volume "Experiments I" Richard Ernst planned the Nobel Prize winning experiments on the application of the Fourier transform and recorded them in detail.

The first time I emerged from the lab carrying a heap of punched cards with my NMR data, my colleagues seemed slightly amused by what they saw. By using my method, I had managed to collect the measurements in a matter of milliseconds. For comparison, an experiment under the old method took several minutes, but you then had the finished results directly on the paper in front of you. By contrast, it took me days to analyze my data, sometimes weeks depending on whether I had access to a computer straightaway. At the time, we were not convinced that this technique could in future save time and improve the sensitivity of our measuring instruments.

I made my breakthrough in the summer of 1964. Just at this time, Wes Anderson was on a business trip abroad, touring South America to make contacts in the scientific community, who may go on to be potential customers for our NMR equipment. When he returned home, I was able to show him some positive results for the first time. I had noted down the design and procedure for the experiments by hand in my lab journal "Experiments I"; I had carefully pasted in the resulting spectrums, which looked like the sort of heartbeat curve you might expect from an electrocardiogram. The clear plots with the definite signal peaks – and exactly where you would expect them – were as much of a surprise for me as they were for Wes.

Later on in 1966, we were lucky enough to take delivery of a PDP-8 minicomputer with 4,000 bytes of memory, which we used to run our Fourier transform calculations in a matter of minutes. This meant that we were already able to arrive at our results as quickly as we did using the old NMR method. We wrote the computer programs we needed for this ourselves using the primitive programming languages that we had previously had to learn. Slowly but surely, the whole NMR department at Varian began to be caught up in our "Fourier fever". We built better amplifiers and busied ourselves refining the computer programs. In just a short time, Varian was able to take out patents covering these new NMR methods. Even today, most NMR devices are still based on these methods. Indeed the process used to produce medical images – magnetic resonance imaging (MRI) – would not have been possible were it not for the pulsed radiation of radio waves and subsequent Fourier transform; there would have been no revolution.

Our intention was of course to publish our results so we sent the paper to the *Journal of Chemical Physics*, which rejected our article twice, claiming it was overly technical and had nothing to do with chemistry. We then tried our luck with a specialist publication, the

Review of Scientific Instruments, which finally accepted and then published our results following a further revision in autumn 1965. The initial response from the scientific community was less than euphoric, to say the least. Many people claimed that the Fourier transform made little sense when applied to physics; others said that the recalculation was pointless given that no new information was obtained that was not already available from the original measurements. The idea that this technique would ultimately prevail because it increased the sensitivity of the measurement equipment by up to a factor of 1,000 was difficult for many to comprehend at the time.

The American way of life

Shortly after this scientific breakthrough in the summer of 1964, Magdalena gave birth to our first daughter, Anna, on 26 September 1964. "We're completely infatuated with our little girl and can't get enough of her," wrote my wife to her godmother. "She already has a few curls of hair and lovely long eyelashes. And how she wriggles about!" There is a picture of me feeding my little daughter milk pudding, which I would always proudly show people. "Richard set the breakfast table so nicely, with flowers and candles," Magdalena told her godmother after Anna's second birthday in September 1966. "Anna was so happy, shouting 'hatchoo' at the top of her voice. Anything that is colorful is called 'hatchoo': The Christmas trees in the market, the advent crown above the fire and the red apples, mandarins and oranges – and of course the lovely flowers in the garden." It was coming up to Christmas. "There are lots of Christmas parties in America, which we are invited to as a couple. However, we'd prefer a quiet Christmas as a family, without too many people about. Richard is busy in any case (he's writing a book), meaning that it's difficult for him to find time for any of these events."

So everything was going really well for me personally. By "well", I mean that I was free to go about my research as I wished. "It seems that turbulent times tend to come to an end when you arrive in the safe haven of marriage (at least that is the case for us)," continued my wife to her godmother. Although I was working a lot, I was always near to my family given we only lived three streets away from my lab in the Stanford Industrial Park. In those early years, we rented a small house on Kendall Avenue, with a lovely garden, trees as if out of a fairy tale and a huge picket fence, with camellias, blue figs and bright red rowanberries. In the nearby wood, but in our garden as well, there were tall mimosa trees, whose heavy scent would fill the entire neighborhood each spring. My wife was amazed at how geraniums would grow out of the ground. Everything almost seemed unreal – some of it like paradise, but some almost alien. Magdalena decorated our house with all sorts of things we had brought with us from Switzerland and she printed curtains, even for my office; she said I should stare at the pattern on them while I was thinking about my work and needed inspiration. "I really do have a lovely husband," she concluded playfully in one of her many letters she sent home. "He helps me dry the dishes every day!"

We were frequently invited for meals at people's houses and we, too, returned the favor, hosting lots of researchers we had befriended, many of whom were Swiss. One was the pioneering computer scientist Niklaus Wirth, whose children loved to play with ours in our garden. Like me, Niklaus came from Winterthur, was a similar age, and was a researcher in Palo Alto at the same time as me. He was an assistant professor at Stanford University in their computer science department. He would also later return to ETH and became known as the inventor of the Pascal programming language. There were also other Swiss researchers working at Varian and we would often all meet up together for garden parties to celebrate Swiss National Day on 1 August or at Christmas time. It sometimes seemed like we had landed on an island populated entirely by Swiss people. "American life hasn't managed to have that big an influence on our lives thus far," wrote my wife, "or perhaps it's just that our sense of 'Swissness' is that much stronger!"

But even Magdalena, who at first thought America to be superficial and full of "plastic smiles", was starting to rethink her views: "I really have to admire the straightforward nature of Americans – how open they are to new ideas – their open-mindedness and impartiality. Our starting point is usually that something is impossible or incorrect. Americans will only think that if they have actual proof that it is. Of course we still find ourselves laughing or shaking our heads at a lot of American things, for example their addiction to vitamins, their fear of bacteria and infections, their worries about leaving children on their own for just 15 minutes, or their fear of crossing a six-lane highway on foot, even though they do this many times a day in their car." As a mother, she experienced the differences between America and Switzerland more clearly than I did. She had to take Anna to the pediatrician every month after her birth to get the necessary vaccinations. "It's interesting to see the differences when it comes to childcare," wrote Magdalena. She was amazed how little American women tended to breastfeed their children and how babies were often placed on their stomachs because this meant they screamed less.

There were certain aspects of our life in the land of opportunity that we could never truly get along with. We missed Europe's cultural cities and did not like the characterless suburbs and residential areas. The contemporary art we saw in the museums we frequently visited was not to our taste. "Most of it is over-engineered art – very rarely appealing in any way – it's really not for us," wrote Magdalena to her godmother. "Perhaps the country is too big, too monotonous across huge swathes of it – the Americans too restless, the advertising so corrosive, the TV the radio...".

The thing we liked most about the US was its outstandingly beautiful natural landscapes. Just before Anna turned one, we all drove together to Death Valley in February 1965. The way my wife reported our trip was indicative of how we felt in our lives at the time, and also the American way of life we had somewhat adopted: "Our trip to Death Valley was absolutely fine for Anna (you could never imagine doing something like this in Europe!). She slept most of the time; it was only on a particularly rough path in a narrow canyon in Death Valley that there were any tears. But apart from that, the roads through the desert are in such good condition. We stayed in motel rooms each night, with their own kitchens and parking spaces, and could even take our sterilizer with us. We warmed up the bottle with a bottle warmer that you connected to the cigarette lighter in the car. We had disposable nappies and Anna even came with us into the restaurants in the evenings. People here are very understanding of people with children and everything is very child-friendly. In fact our friends here actually encouraged us to take Anna with us to Death Valley, saying it's a great place for babies in winter!"

I later wrote about our holiday in Mexico to my godson Martin, the son of my sister Verena, who read out my letters to him. "We've had lots of visitors recently and we always take the opportunity to show them the 300 slides from our trip to Mexico (last time one of our guests even fell asleep!). That sort of trip showed just how sterile life here in America is, with its uniform customs and traditions. In Mexico, you feel like you have been taken back to the middle ages, with all the striking characters you see, the beggars, the street traders, the mothers carrying their children on their backs, the shoeshine boys, and also the rich. Life here is so good, and is all the richer for its highs and lows."

Overcoming my fear of public speaking

This calming, restorative way of life was interrupted each year by the annual Experimental NMR Conferences (ENC). These conferences, which first took place in 1960, quickly became the most important and influential gathering of researchers who worked in the field of nuclear magnetic resonance. Held over several days, the event grew quickly and nowadays over 1,000 researchers from around the world still convene here each spring – a blend of eager students, ambitious PhD students, self-assured postdocs, assistant professors, the confident "old hands" in the field, but also a number of engineers, technicians and salespeople from various manufacturers of measurement equipment. Through 60 or so acclaimed lectures as well as hundreds of posters presented in the congress center's exhibition halls, we would come together to learn about and discuss the latest and most interesting innovations in research. The list of after-dinner speakers was basically a "who's who" of the history of NMR.

For me, these conferences were the highlight of the year. In the first few years, they always took place in Pittsburgh, which saw me board a flight each March or April to spend a few days in the eastern US. I delivered my first lecture in 1964 at the fifth Experimental NMR Conference. A year later, I returned to present our breakthrough in "improving sensitivity through Fourier transform technologies". I can no longer remember whether the presentation was met with as much interest as – with hindsight – it deserved. In any case, I was always meticulous in preparing my lectures and would spend two or three weeks doing so; nevertheless I would still regularly worry that they just didn't have enough content. So I would dig out all the notes that I had stuffed in my drawers and hoped that this would at least mean there was something, new, original, or even spectacular in my lectures. I would often not only talk about projects that were already finished or my results from the lab, but also ideas that I was considering for the future.

The leading NMR researchers at a scientific conference in Tilton, New Hampshire, in the early summer of 1965. Richard Ernst can be seen in the fourth row, second from the right.

I suspect that the patent lawyers at the companies I worked for often looked on at my openness in this regard with a sense of desperation. Our field had long since become a business that generated a lot of patents and in which there was a lot of money involved. This soon gave rise to a curious mix of researchers openly exchanging their thoughts and ideas with each other, all the while amidst an air of secrecy and a relentless race to be the first to make a given discovery. Quite a few speakers were therefore reluctant to talk about their ideas, lest they give them away. None of this bothered me, however. I never kept quiet about my ideas out of fear that somebody would beat me to them or that it might have a negative impact on the "novelty value" of my inventions. Looking back, I can't think of an instance in which there were negative consequences of me sharing an idea with others, an idea that others would have most likely buried in secret documents somewhere.

I worked very hard on the presentation of these lectures, drawing all the figures and slides myself. At first, I would use scissors to cut up the colorful slides; later on I used the latest technology available – I had no trouble moving from film overwraps to slides and then ultimately to PowerPoint presentations. I always used a great many slides and pictures, because as the phrase goes, a picture is worth a thousand words. Given that public speaking was never one of my strengths, I considered each of these "thousand words" to be a thousand words well saved on my part. Nevertheless, I paid a lot of attention to the language I used, always holding the lecture in the present tense. I also managed to include a few witty lines to keep the audience amused. It was often the case that funny turns of phrase or little jokes that spontaneously occurred to me during the lecture got the most laughs. This acted as a real motivation and the promise of a warm applause at the end always spurred me on. But I would always live in fear of not living up to my own expectations; even in the 1990s, I would still record in minute detail how my lectures were received by the audience. My diary entries would contain comments such as "successful lecture", "average", "wrong audience", or "just wasn't captivating".

I was never able to quite get rid of my nerves and worries before an important presentation. When I returned to ETH, I dreaded having to deliver my first lectures to students. It transpired that these lectures were absolute disasters, with students sat there laughing and whispering among themselves. One student later confided in me that they simply didn't take me seriously. However, I intuitively realized that being able to convey my ideas – whether at scientific conferences or in front of university students – was the only opportunity I had to distinguish myself as a scientist. And in actual fact, my lectures were gradually becoming more successful.

The speeches I gave were attracting more praise, with audiences finding them informative, innovative and humorous. I began to notice that at scientific conferences, the audience would flood into the room when it was my turn to speak. The organizing committee of the ENC once carried out a survey to find out which presentations were the most popular. The results flattered me, saying that I could often have filled the room twice over! And the rule that the same speaker should not appear in two consecutive years so as not to bore the audience appeared not to apply to me: Between 1976 and 1987 (with the exception of 1985), I could be found somewhere on the list of speakers.

Richard Ernst is holding his daughter Katharina, who is just a few weeks old, and Anna, who was born in 1964, is sitting next to him. Photo taken in summer 1967.

On 26 May 1967, Magdalena gave birth to our second daughter, Katharina, in Palo Alto. By this time, we had already decided that we wanted to return to Switzerland before our children started school. "The way children are brought up in America is not one of things we like here," wrote Magdalena in a letter home. "By that I mean it is more the mentality of the parents that influences how they are brought up. [...] Here, many people are so empty on the inside, even if their table is bending under the weight of all the fine food on it, even if modern devices do all their work for them, even if they can have everything their heart desires. It sometimes feels outrageous living in a country like this. You find yourself asking where this will all end up." Our days in the US were numbered. Less than a year later, in March 1968, we returned home to Switzerland after having spent five of the happiest years of our lives in sunny California. Had I anticipated the difficult times that lay ahead of me, perhaps we would have made a different decision.

Magdalena Ernst in the doorway to the garden of her house in Palo Alto, around 1967.

Return to ETH, 1968–1990

A hard landing in Switzerland

Following five productive and inspiring years in California, my return to Zurich in spring 1968 felt like being thrown back into the Dark Ages. I had nothing I needed. I felt isolated at ETH, I had no support and the equipment I had at my disposal was antiquated and totally unfit for my work. The one ray of light was my frequent exchange of letters with Wes Anderson. In a message sent just two weeks after arriving back at the end of April, I pulled no punches in letting him know how bad things were at ETH: "I feel that there is a good challenge to try to form a productive NMR group," I wrote. "One feels pretty much isolated and does not know where to start, it is completely different from the start at Varian. One is not even introduced to anybody and does not know what the neighbour is doing, except that he is building some kind of a complicated instrument and probably does not know what to use it for."

There were various reasons for my returning to the university, despite having sworn just five years previously that I would never set foot in it again. Firstly, the situation at Varian Associates had changed. I was increasingly being given work that I had little interest in doing and the new decision-makers in the company had not recognized the potential of our Fourier transform method. They had failed to see that the method produced results far more quickly than the slow scanning method from the Felix Bloch era, or that being able to take faster measurements automatically meant greater sensitivity in NMR methods. It was more efficient all round: You needed less material for the sample and the base magnetic field did not need to be as strong in the experiment, which reduced electricity costs. The project would have been a real cash cow. While Varian patented the invention in the US at the end of October 1969 as Anderson-Ernst patent no. 3.475.680 (for which I even received a small bonus of around 100 dollars), they did not push forward with research in this area. Although Wes and I had produced some extremely promising results, it would have taken a lot more research and development to be commercially viable. Rather tellingly, Varian was not the first company to consistently apply our methods in a spectrometer; instead, it was the German-Swiss firm Bruker, which in 1969 began to develop a new spectrometer concept based on the Fourier transform.

But the manager at Varian had something different in mind for me: I should either focus on projects that would quickly produce commercially viable products or I could become a service technician for existing product lines. Neither of these options appealed to me. I was never that interested in the commercial aspect of our work, it just wasn't in my nature. I much preferred perfecting and developing measurement equipment rather than simply using it. And I wanted to pursue a career in research. Each success simply whetted my appetite further.

Considerations regarding our family also played a part in our decision to return to Switzerland: Magdalena did not want our children to go to school in the US – "Having Mickey Mouse in the children's room is not to my taste," she would say at the time. Our American friends had some trouble understanding our decision. They were confused as to why we would voluntarily swap the exciting, progressive atmosphere in Silicon Valley for the "backward-looking" Old Continent.

It was then in 1967 that my PhD supervisor Hans Heinrich Günthard got in touch, offering me a position as a research assistant at the ETH Laboratory of Physical Chemistry. Following my previous departure and Hans Primas' change of direction towards theoretical chemistry, NMR research there had been abandoned. However, now Günthard wanted to bring this field back to life, primarily as a means to complete his institute's portfolio of modern analysis methods based on the fundamentals of physics.

He definitely had a good nose in this regard. Thanks to their precision and ability to provide meaningful results, these physical methods had come to dominate chemistry since the Second World War and there had been a sea change in the way they could be applied. As Günthard himself had specialized increasingly in optical spectroscopy, however, he needed to find the right people to work on NMR methods – and so he recruited me. I knew Günthard and appreciated that working together with him on a constructive basis could be difficult, so the idea of returning to work with him was disconcerting to say the least. But ultimately, I had no real choice. I had since become so specialized in my field that there were perhaps just a handful of jobs in the world that would be suitable for me. So I accepted Günthard's job offer even though he was "only" taking me on as a research assistant, his "underling".

Richard Ernst at the age of about 36, photographed in Winterthur. After returning from the USA, he and his young family moved back into his mother's house.

When I rejoined ETH, I had very few resources at my disposal – either in terms of money or staff, both of which I urgently needed to properly continue my research. I was given a small office in the new departmental building on Universitätsstrasse. I had the impression that my achievements in California barely registered with my peers in the Old World. I had come back with some outstanding results and solid publications under my belt, and had performed numerous theoretically sound experiments and even developed innovative computer applications to accompany them. It would be fair to say that I had brought NMR methods into the digital age. I had reasonably expected that I – a leading light in NMR methodology – would have been inundated with invitations to conferences, universities, companies in the industry, and so forth. But I got nothing of the sort. My work had either been ignored, or there was simply no interest in innovative research.

I was given sole responsibility for operating the various NMR equipment. In addition to my research work, I was placed on 24-hour call-out for chemists wanting to analyze their synthetic substances in the lab as quickly as possible. However, we only had a smaller spectrometer, which was mainly for students to practice with. Apart from that, there were the old devices lying around that Hans Primas had built many years previously. "Nobody took yet a meaningful spectrum on it," I wrote to Wes, "there are no data about sensitivity or resolution available." I continued: "Unfortunately, there is not even an electronics technician around, so I will have to do some down-to-earth electronics myself." In light of this, it was practically impossible to continue my own research. While Günthard considered getting the lab a new, more powerful 220 megahertz spectrometer with superconducting magnets, money was tight. And even this device would have been completely unsuitable for my work.

I felt like I was being deliberately obstructed. The atmosphere was one of jealousy, with the fight for recognition and resources constantly bubbling under the surface. My view of Hans Heinrich Günthard went further downhill, and I became ever more critical of him. He would not tolerate having people around him who attempted to compete with him or question his authority. He was driven solely by objectivity and scientific thought, always emphasizing the rational motivation behind all his decisions and insisting that his own personal views were of no importance. However, he would often react to things in a completely irrational way and make decisions that nobody could understand. He would grant his favor like a monarch, working really well together with his closest staff, all the while refusing to accept any comments, ideas and certainly not criticism from "bad colleagues" like me. "We've known that for ages already" or "well that's obvious", he would always respond when I tried to contribute a new idea. As his "servant", I was there to do whatever he wanted and he was there to reap the benefits of his colleagues' work. Hans Heinrich Günthard was a strange man, who never let me see behind his mask. He was never interested in personal matters; all that mattered to him was results. The person behind the research was of no interest to him. When I – his "employee" – won the Nobel Prize, he briefly acknowledged this and then quickly moved onto the next item on his agenda.

After six months, I was given a lab technician to help me; "not the greatest, but he'll probably get better," I wrote to Wes in November 1968. Finding a capable programmer was still impossible because – with the exception of the experts in the Institute of Applied Mathematics – there was almost nobody at ETH who had any real knowledge of computers, knowledge that was key to making progress in our field. "People here are only just discovering the benefits of computers," I wrote in one of my letters back to Silicon Valley, where scientists had long been relying on digitalization to make progress in their fields.

My area of research also had a problem insomuch as it was struggling to attract up-and-coming scientists. I complained to Wes that "there's barely any new students joining our department; people are scared of the mathematics that the subject requires." However, I felt partly responsible for this myself as starting in autumn 1968, I too had to deliver lectures to students. I had submitted my habilitation thesis while I was still in California and had finally become an outside lecturer. However, the courses that I then had to teach undergraduate students were somewhat daunting. I was still yet to overcome my fear of speaking in front of people and I spent many a sleepless night worrying about making a fool out of myself in front of students.

For better or worse, I just got on with the difficult situation at ETH. Instead of discussing my ideas with colleagues at ETH, I instead did so with Wes Anderson across the Atlantic via airmail. We sounded out the potential of the Fourier transform, considered new ways to excite the magnetic nuclei and discussed all the potential methods there were to use NMR spectroscopy both in theory and practice. Wes would frequently ask me questions about my ideas and the projects I was working on; in turn, I would report to him on both my successes and failures and the work being undertaken by groups I was close to. I wasn't just friends with Wes – on returning from California I had agreed to continue acting as a consultant for Varian Associates, for which I received a small paycheck each month. I was not always able to meet these expectations as months could go by without my being able to complete any useful experiments. This burden weighed heavily on my shoulders and must have been apparent in my letters to Wes, even though they nominally dealt with scientific questions. This is the only reason I can think of for him sending me the following request in January 1969: "Richard, we all miss you very much here at Varian," wrote Wes, "and I would like to ask you to seriously consider returning. We could offer you better conditions, more freedom and responsibility, salary increase from your previously salary, and the department managership of Forrest Nelson's old department. Please let me hear your response to this proposal."

I took my time to think over his offer and it was all of six months later at the end of July 1969 – by which point we had since exchanged several other letters dealing mainly with scientific matters – that I answered his question: "I never answered your question about coming back to Palo Alto. We finally decided to stay in Switzerland [...] We certainly do not have it better here than we had it in Palo Alto, but we do not have it so much worse that we could decide to leave our home country. Just in the moment, we have to stay, otherwise I probably would lose my job. It is really very hard to advance here. I do not know, when I will get an assistant professorship only. With my very best regards, yours, Richard." I was struggling with the fact that I was yet to be offered a permanent position at ETH – after all, I had a family to feed.

Then in autumn 1969, I was awarded the Ružička Prize, a prestigious award for young researchers, which is still presented today. I was proud to finally see my name among the impressive list of previous winners. I had worked towards this honor while completing my dissertation, but had never quite got there, but now my time had come. I was awarded the prize for my "comprehensive and in-depth analysis of the problems of increasing sensitivity" in the area of nuclear magnetic resonance. This was based on my work and publications about the Fourier transform, which I produced together with Wes in California.

Unfortunately, there was a cloud hanging over this success as I had a guilty conscience with regards to my friend Wes. It took me six months to write to him to tell him about this "unexpected" honor that had been bestowed upon me and even then, I felt embarrassed. "I am only telling you," I wrote, "because I have you to thank for a lot of what was said in the speech about my work, especially the work on improving sensitivity and the Fourier transform spectroscopy." This was not a problem for Wes, ever the generous type: "You are right to be proud of the huge amount of work you put in to improving sensitivity and the Fourier transform spectroscopy. I'd like to congratulate you for your excellent work."

In reality, however, I was simply getting nothing done at ETH. Theoretically, I had a great many ideas for experiments, but in practice it was impossible to try them out. To do so would have required properly designed spectrometers, more powerful computers and above all, help from a programmer. So I had to write most of the programs myself. "I still don't have a good example of a linear measurement," I complained to Wes in September 1969. "I'm still having difficulty implementing the fast Fourier transform. The interpolation isn't working." And so it went on. The most annoying thing was that the measurements generated by the NMR spectrometers I had were constantly being interfered with by the trams passing by in the street outside. The electric tram heading towards Irchel in northern Zurich ran right past our lab. Every time a tram passed by, the stray radiation from the tram lines massively interfered with our magnetic field. To obtain a precise measurement, a stable magnetic field is a must. An analysis showed that the electromagnetic stray field from a tram could be detected several hundred meters away. I wrote to Wes complaining that there was not a single location at ETH where we could conduct our experiments without our spectrometers being disturbed and I included a small sketch illustrating just how much trams interfered with our readings.

The only way to obtain reasonable measurements was to conduct our experiments between half past midnight and four o'clock in the morning when the trams were not running – hardly the ideal time to be at work.

27 July 1969

Dr. W.A. Anderson
Director of Research
Anal.Instr.Div.
Varian Associates
Palo Alto, Calif. 94303

Dear Wes,

I am sure you can imagine how bad a consience I have not to have written to you more frequently. I do not want to excuse it with my busyness, but it is one of the reasons.

One of the major things which kept me busy for the last few months was the HR220. There was and is still a lot to be done to keep it running. I do not know whether you heard of our problems. At first, the helium losses were very high, about 10 liters per day. We heated it up twice and cooled it down again. The second time, it started to work fine and we do not have further problems since then with helium losses. The strange fact is only that we have no idea what the reason was. There is only one remaining weak point connected with the instrument, its sensitivity is 50/1 which is half of what it should be.

But the most disturbing effect is the interaction between spectrometer and the nearby street-car which works with 500 Volt dc. There is a very strong stray-field predominantly along the axis of our solenoid:

In the moment, we can make high-resolution measurements only from 0030 to 0400, it is not a very agreable time to work. The stray-field amounts to about 1 mGauss which causes line shifts of about 4 Hz with

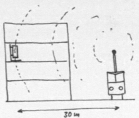

Letter to Wes Anderson dated July 27, 1969. With a sketch, Richard Ernst illustrated how the Zurich streetcars disturbed the NMR measurements in his laboratory at the ETH Zurich.

Kurt Wüthrich on his appointment as assistant professor at ETH Zurich in 1972. The 2002 Nobel Prize winner came to ETH in 1969, one and a half years after Richard Ernst.

On top of all this, office space was becoming a rare commodity. Eighteen months after I arrived back at ETH, a young, confident researcher named Kurt Wüthrich also returned from the US. He had accepted a job as a chemistry researcher at the newly founded

Laboratory for Molecular Biology, which later became the Institute of Molecular Biology and Biophysics. Even before he arrived, my new "old" boss Hans Heinrich Günthard had asked me whether Kurt Wüthrich could squeeze in alongside us in our offices, which given the already tight working conditions, I was less than impressed with. "I'll make it work somehow," was my less than enthusiastic reply, before nevertheless adding that "it's great that he's coming to ETH!" So in those first months, we literally worked closely side-by-side, both of us hidden away in converted temporary offices on the roof of the physical chemistry building. But there were soon disagreements over how research funding and resources should be allocated, and over access to computer time and measuring equipment, which put a real strain on me as the person responsible for the pool of equipment. The situation was diffused somewhat when Kurt Wüthrich moved to the new building on the Hönggerberg a little more than a year after he came to ETH.

A shot across the bows

By this point, I was 36 years old and it seemed that I would fail to achieve any of my scientific goals. Having to take responsibility for the inadequate pool of equipment was the source of many a sleepless night. I felt overwhelmed and impeded, and it took me back to the feeling I once had when I was completing my ill-fated military service. I was undergoing my officer training and was ranked as a sort of "assistant commander". We were on a major operational exercise. It was cold, raining, and getting dark in a small village in the canton of Aargau. I was responsible for more than fifty recruits and was tasked with finding them emergency overnight accommodation as well as a camouflaged shelter to protect our vehicles from supposed air attacks. But I didn't know the village at all and had no time beforehand for any reconnaissance. All I had was a jeep and a driver. I had to go from house to house, ringing on doorbells. The homeowners were dismissive and unfriendly; the cattle farmers in particular were miserly and didn't want to be disturbed, even though their farmyards would have been ideal for our needs. The freezing recruits sat in the troop carriers were growing inpatient; their drivers were waiting for my commands and instructions. So I issued my commands, but these just clearly confused them far more than they helped. What followed can best be described as chaos. Nobody listened to my orders and even I myself no longer knew how to get things back on track. I was completely overwhelmed. I felt like a foreigner in enemy territory, desperate, stressed, with no way out.

I had a similar feeling during those first two years after my return to ETH. Then finally, at the end of April 1970, a ray of light! Wes and I had planned to meet up; after a two year absence, I was once again heading to the most important convention in my field, the eleventh Experimental NMR Conference in Pittsburgh. These meetings, attended by the crème de la crème of the NMR research community since the start of the 1960s, were always a happy highlight of my life as a researcher. This time I had even been asked to appear as the opening speaker. I intended to take them up on their offer and had also planned a trip to Palo Alto before the conference to visit Wes and his wife Jeannette. We had come up with a great schedule: I had agreed with my former boss and friend that we would visit Varian Associates and a number of other innovative labs in Silicon Valley to see the latest projects. I was really looking forward to seeing all my old colleagues, but fate had other ideas. "Dear Wes," I wrote in a short letter dated 4 April 1970, "Unfortunately, I have to tell you that some of my plans have been ruined by a nervous breakdown which I suffered about a week ago. [...] I am sorry for the sad news. With best regards, Richard R. Ernst." What had happened? What was the reason for me sending such a letter? How does a scientist who prefers to bottle everything up inside rather than give away any inkling of failure find himself admitting to defeat? What would cause a researcher who was previously willing to sacrifice everything for his work to backtrack in this way?

Everything had become too much. It was shortly before the Easter holidays, on Good Friday or Easter Saturday perhaps. We were expecting a couple of guests on the Sunday, with whom we would celebrate Easter, when I suddenly collapsed and lay unconscious on the floor. Magdalena found me and in shock, hurried to me, but I lay there still for 30 seconds, perhaps a minute or longer. She phoned a doctor straightaway, who determined that I had suffered a nervous breakdown. He prescribed me a period of rest and told me I was not allowed to go to California. I sat around dejectedly, with no interest in anything; I was completely apathetic the whole weekend. We had to tell the guests we had invited for Easter that we were having to cancel our plans. I thought that this would be the definite end to my scientific career. I envisaged myself spending the rest of my life stamping letters behind the counter of a post office.

We – just Magdalena and I – were able to get away for a couple of days to Ticino in the south of Switzerland. We knew that our two daughters Anna and Katharina were in good hands at my mother's house and a generous friend of ours had kindly let us use his holiday house in Arosio for several days. It was already sunny and warm, and we spent our time walking alongside the lake or discovering the pretty old town in Lugano. Here we indulged our passion of rummaging through junk shops and antique dealers, on the lookout for hidden gems and other artistic treasures. It was in a cuckoo clock shop on the Via Nassa that we discovered a type of Tibetan painting – it was a beautiful scroll painting, also known as a thangka. We had come across these mystical paintings, the thangkas, for the first time almost exactly two years previously to the day, when we were returning home from California. Magdalena and I had decided to make a detour for a few weeks to Asia before arriving back in Europe. We were in Kathmandu when we saw them for the first time and I had not been able to stop thinking about them since. This part of my story is so important to me that I will talk about them in greater detail in a later chapter. But now, stood in this antique shop in Lugano in my miserable state, this passion was like a lifeline in my hour of need. The aura this painting had fascinated me from the outset and on this occasion, my love of collecting art overcame me.

Music too helped me through this difficult time. I would listen to a great many contemporary composers, from Igor Stravinsky to Karl Amadeus Hartmann, Charles Ives through to the enchanting symphonies of Dmitri Shostakovich. Classical music had become the medium in which I felt truly at ease, whether as a listener or as a musician playing my cello. I had become enthralled by it during my early school years when my childlike aspirations had focused on the glamorous appearances of the many international starts at the Musikkollegium in Winterthur, and now it was offering me an escape to a world in which feelings and thoughts could be expressed without using words.

These experiences had taught me that even scientists need more than one leg to stand on in order to move forwards. Slowly but surely, I managed to get back on my feet – I'm still proud of this today. Nowadays, people would probably go to see a psychotherapist or counselor, but that was out of the question for me. This simply didn't fit with my image of what a successful, ambitious scientist should do. We from the Ernst family simply don't visit psychiatrists!

By the end of April 1970, I was even fit enough to fly out to the convention in Pittsburgh and on 22 April, I gave the opening speech. It was a real success judging by the reaction of the audience. The session was being chaired that day by Tom Flautt, who still remembered my speech years later: "I remember this paper for its clear mathematical explanation of the difference between spectroscopy using a single frequency and multiple frequency excitation produced by random pulses, which results in improved signal/noise," he wrote in an article for a highlights publication of all the ENC conferences. While the problems at ETH Zurich had obviously not disappeared overnight, I was now able to approach them with a calmer mind.

Jean Jeener's advance into the second dimension

In 1971, I had four or five PhD students, with whom I could gradually press forward with the experiments and approaches I wanted. I had now been an assistant professor in the Laboratory of Physical Chemistry for more than a year and an associate professorship was within my reach. Although our pool of equipment was still outdated and did not come close to meeting the highest standards at the time, we diligently improved our methods, invested in innovative computer applications, refined the Fourier transform method and tried out various excitation systems, that is to say the various options for how radio waves could be fired at samples to excite their nuclear spins. Even though we did not make any revolutionary breakthroughs, we did learn how to use a whole range of NMR methods – both theoretical and practical.

From 1970, the Fourier transform method was also used in measuring equipment. Thanks to the increased sensitivity it offered, it was possible to analyze ever more complicated substances – right up to small proteins. However, complex spectra resulted from the up to 500 magnetic hydrogen nuclei in different chemical environments that make up these molecules. As a reminder, a molecule contains various atoms, which are connected with each other in different ways Most organic substances contain only a few elements: Carbon, oxygen, and many hydrogen atoms, perhaps with a nitrogen or sulfur atom there too. Not all atoms are magnetic, that is to say they have nuclear spin. In practice, these are mainly the hydrogen atoms – and it is these atoms that nuclear magnetic resonance methods tend to observe.

Richard Ernst and his doctoral student Thomas Baumann discuss the novel 2-D NMR spectroscopy, a revolutionary method, of which Baumann became aware at a summer seminar in 1971.

Notably, within a given chemical substance, the hydrogen atoms are in a different local environment. They might be bound to a carbon atom, for example, with others bound to an oxygen atom. While these neighbors do not have nuclear spin, they do have electromagnetic properties, which have an impact on the magnetic moment of the hydrogen atom at an atomic level. The result of this is that you can use the NMR method to measure minimal shifts in the resonance frequencies depending on whether a hydrogen atom is bound to a carbon or an oxygen atom (or indeed another atom). The Fourier transform method made it possible to distinguish between ever smaller differences in a spectrum. This means that signals from an ever increasing number of different hydrogen atoms could be deciphered, and that it was suddenly possible to identify differences that would previously have gone unseen. In other words, the "spies" inside the molecule – which let us know where they are, who they are sitting next to, and how they are bound to each other – were doing better work. However, this technical progress also threw up new questions. Just how could we go about examining the incomparably more complex molecular structures that occur in nature? How could we identify and study the much larger proteins, genes, and many other important components that we find in the living world?

It was in autumn 1971 when the stars unexpectedly aligned. One of my first PhD students, Thomas Baumann, returned from a summer school for young researchers held in what was then Yugoslavia with an invaluable scientific discovery. He had carefully taken notes in a lecture held by a Belgian physicist, whom I had previously known by name only – Jean Jeener, who was then a researcher at Brussels University. He was in Baško Polje, a lovely resort on the Croatian Adriatic coast, to give a lecture to the aspiring young scientists about a new technique he had invented called two-dimensional NMR spectroscopy. At the time, it seemed almost esoteric. The hydrogen nuclei spinning in the magnetic field are exposed to two consecutive radio wave pulses and their response is measured after each pulse. In technical terms, only the signal after the second pulse is recorded and you can observe the way in which this signal is influenced by the delay between the first and second pulse. However, the experiment is repeated multiple times with carefully chosen time steps. The first pulse excites the nuclear spins as usual, while the second disrupts them briefly. By analyzing and comparing the signals, you can create an informative two-dimensional spectrum, with one of the dimensions relating to the time between the pulses, and the other corresponding to the time elapsed after the second pulse. The key aspect of the technique is that the way in which the nuclear spins react to two of these pulses provides more information about the state or position of the nuclei in the molecule than if they had only been exposed to one pulse.

Because this experiment is so key to my work, I'd like to explain it using an analogy. Imagine you are on a farm and come across a calm, quiet flock of chickens. If you were to then fire a gun while standing next to them, the noise would scare them and they would all fly up in the air. Shortly afterwards, if you were then to fire a second shot, you could observe where and when the chickens would return to pecking around on the floor. The time between the first and second shots would obviously have a significant influence on where the chickens would land. If the second shot is fired straight after the first, it's possible that lots of the chickens would still be in the air and couldn't flutter any more. If there is a longer delay between the shots, the faster birds would already be further away and have settled down.

The special thing about two-dimensional NMR spectroscopy is now that while you can only observe the "flock" of nuclear spins after the second shot is fired, you can see their reaction and draw a conclusion as to what happened between the shots. And if you then fire a second, third or even fourth shot, you can increase the number of measurement dimensions as you wish and thus obtain even more information from them.

I like to use another comparison to highlight the advantages of a two-dimensional spectrum over its one-dimensional counterpart: Imagine the silhouette of some mountains in the morning fog, as you might see on a golden autumn morning in the Swiss Alps. They appear to you like a long row of peaks with different heights and if you use the right app on your phone, it is not hard to work out the heights of the individual mountains. Now imagine the same mountain peaks, but this time photographed from directly above using an aerial camera at 10,000 m. It goes without saying that you would be able to quickly glean more information from the panorama recorded than simply the heights of the mountains. You would see all the topography of the mountain range, the deep valleys, the distances between the peaks and where the ridges run.

Back in Zurich, we discussed Jean Jeener's approach in our mandatory group seminar, which I always held whenever one of my PhD students returned from a conference or training seminar. Few of the students there understood the point of the exercise, but I was enthralled by the idea of a two-dimensional NMR spectrum. Everything seemed to fit together perfectly. It was as if I was listening to an astonishing symphony, as I used to do as a child at the concert rehearsals at the Musikkollegium in Winterthur. The polyphonic melodies meshed together harmoniously (or disharmoniously) and took listeners into another world without them knowing how exactly. The sounds made by the various instruments followed one after the other, overlapping, amplifying or muting each other, swelling once more as the musicians played another note and then dying away over time – this continuing, harmonic change from forte, crescendo, diminuendo, piano, pianissimo... I could imagine how something similar would happen in the art of nuclear magnetic resonance. The nuclear spins react when they are stimulated, they get faster when exposed to a pulse, and slow down following a second pulse; many by their nature make a dull sound, others have a clearer sound, and there are some that can barely be heard at all. Stemming from my passion for classical music, I had a good feeling for harmonies and disharmonies, for wave-like and decaying sounds and frequencies that could be observed during the course of an experiment, all of which helped me greatly in understanding the world of nuclear magnetic resonance and developing Jean Jeener's idea.

This was exactly the technology that I had been waiting for. By generating a spectrum with this structure, combined with the rapid Fourier analysis, it was possible to obtain far more information about atoms than was previously the case. Putting the technique into practice was of course somewhat more difficult. I had previously considered experiments involving multiple pulses, but I had great respect for just how complex the results would be. Jean Jeener had himself conducted initial experiments, but the resulting spectrums did not provide any advantages over the previous methods, so he didn't pursue the project. I was also hesitant about continuing down this path, as I was reluctant to tread on his toes. I nevertheless had one of my most talented PhD students, Enrico Bartholdi, at least run through the theory behind the idea. The results on paper were a revelation. It seemed to work, but there was more: It was like we had been able to open the door a crack and behind it was a real Garden of Eden for NMR spectroscopy. We suddenly had ideas for thousands of experiments using different pulse sequences, which would provide completely new information. I had never been so excited in my scientific career before. But we held back with experimenting for a time.

Jean Jeener (left) in conversation with Luciano Müller, another doctoral student of Richard Ernst. It was the Belgian physicist Jeener, who presented the idea at the summer seminar. It took NMR technology in a new direction.

"Pulse pair technique in high-resolution NMR", the speech given by Jean Jeener at the Baško Polje summer school on 13 September 1971, has since become a thing of legend, even though the work was unfortunately never published. The fact that the summer school took place during the Cold War in the then non-aligned, but nevertheless communist Yugoslavia is particularly relevant as it says a lot about the independence of scientists at the time. The conference was organized by the Groupement AMPERE, which was founded by French scientists in 1951 and is worth mentioning. The society was founded as continental Europe's response to the research being carried out in the English-speaking world in the field of microwave technology. The name "AMPERE" stands for "Atomes et Molécules par Études Radio-Électriques", that is to say research into atoms and molecules by studying radio waves. The idea was to push forward research in this area by exchanging ideas and working together, but one of the society's most important goals from its inception was to foster scientific and social understanding between east and west.

In general, relations with scientists in the Eastern Bloc were good, even during the Cold War. There was a real hunger for knowledge behind the Iron Curtain and scientists there were as talented as their western counterparts. Shortly after I returned from the US, I received an invitation from the Physical Society of East Germany to attend a conference at Karl Marx University in Leipzig. Lectures in Leipzig had to be delivered in German, Russian or English, and I spoke in German on the subject of "methods for simultaneous broadband excitation in high-resolution NMR spectroscopy". Glancing down the list of speakers, you saw primarily researchers from the West, including the later Nobel Prize winner Peter Mansfield and British researcher Raymond Andrew, both of whom made key contributions in developing medical imaging.

Crossing over to the living

I just could not stop thinking about Jean Jeener's idea for two-dimensional nuclear magnetic resonance which he had revealed in 1971 at the summer school in Yugoslavia, but I was hesitant to perform any of my own experiments before Jeener had himself published his own work on this subject. I was torn and so discussed my dilemma with my PhD students: Would it be unethical to not wait an appropriate length of time? But we sat there waiting in vain – Jean Jeener just wasn't publishing anything! Perhaps the concerns I had were a little excessive. After all, we were in constant contact with "Brussels", exchanging scientific ideas and thoughts, always being sure to let Jeener know when we had made any progress. He later told me that he was always pleased to receive news from Zurich!

Then I found an experimental approach that incorporated Jeener's basic idea, but with a completely different goal in mind. This approach took us directly to the process used to produce medical images, which would go on to be so important in the field of medicine. I was inspired by US chemist Paul Lauterbur, who I heard speak at a scientific convention in April 1974 to which I had also been invited. Lauterbur, a pioneer of magnetic resonance imaging, presented the first three-dimensional MRI images of a mouse. This was at half past two on a Tuesday afternoon, and I quickly noted down the techniques he was describing in my notebook. For the time, the picture was extremely good; you could see the internal organs and all the contours, but not the bones, as is typical for MRI images. In the middle of the mouse, however, was a mysterious white spot, which didn't seem to belong to an organ. Was this some sort of ghost that had made its way into the image?

I immediately realized that this white spot was a result of the special process that Lauterbur had used to produce the image. Although an MRI image of an object looks like a photograph, it is no such thing: It is actually the result of NMR spectrums being converted. As is the case in chemical analysis, hydrogen atoms are placed in a magnetic field and then their resonance frequency is measured; here, however, this takes place in the human body. In real terms, this involves measuring the hydrogen atoms contained in water, of which there is more than enough in human bodies. This magnetic resonance method is also known as magnetic resonance tomography (MRT) and is a tomographic method used to obtain cross-sectional views of the human body. The sectional plane of an image produced in this way runs along the magnetic field; across this "slice", all the hydrogen nuclei are excited and then measured.

But it is not that simple. If you were to apply only a constant magnetic field, you would see nothing more than a white spot, because the hydrogen atoms on the left side of the body have the same resonance frequency as those on the right side. To address this problem, Paul Lauterbur and other scientists had already developed a more refined method. They applied a second field on top of the constant magnetic field and then gradually increased the strength of this second field along the sectional plane, for example from left to right. This resulted in the hydrogen atoms on the weaker side of the magnetic field reacting differently to those on the stronger side; the resonance frequencies become higher from left to right. Here the hydrogen atoms are said to have undergone position encoding; following computer-based analysis of the results, a cross-sectional view can then be produced. At the same time, these pioneering scientists discovered that by performing targeted experiments, it is possible to distinguish between the resonance signals of the hydrogen nuclei depending on the tissue being examined. This created the foundations for its use in humans, making it possible to distinguish between lungs, livers, kidneys and hearts and even to identify cancerous tissues.

To create a three-dimensional image, Paul Lauterbur, who would later go on to receive a Nobel Prize for his work, measured the bodies he was examining on a level-by-level basis, analyzing the resonance frequencies and then bringing them all together to produce an image by applying a reconstruction algorithm. This method was similar to the old NMR method used to analyze chemical substances, when you had to sample one frequency after the other, a laborious, long-winded process. In those days, if you had tried to use it to create images of humans, you would have had to place them in the magnetic field for hours, if not days. Then I got the idea of combining the principles of the Fourier transform with 2D spectroscopy and applying it to the imaging process: Instead of measuring each slightly different level on a step-by-step basis, I intended to use less selectively chosen pulse sequences to excite all the hydrogen nuclei at once. This once again resulted in complex series of signals being produced; now, however, I was able to use the Fourier transform together with ever improving computer programs to unscramble them.

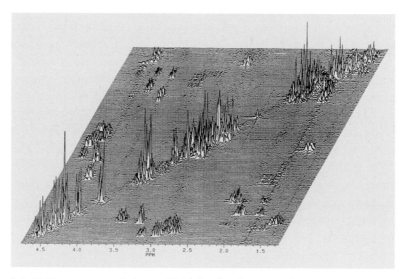

2-D NMR spectrum of the artificial fertility hormone buserelin. The peaks show the spatial relationships of the hydrogen atoms. This makes it possible to determine the 3-D structure of the molecule.

Now I just needed some results – and time was not on my side: In September 1974, some of us Swiss NMR researchers were planning to hold an international scientific convention in Kandersteg in the Bernese Alps. With the sixth International Conference on Magnetic Resonance in Biological Systems, we had managed to attract a major international gathering of NMR researchers to Switzerland. The driving force behind all this was Kurt Wüthrich, who had resolutely pressed ahead with his research in the area of biomolecules. He had invited me and NMR specialist Joachim Seelig from Basel to sit on the organizing committee. Over 220 scientists from 23 countries ultimately made their way to the Bernese Oberland, to a convention at which I very much intended to unveil my spectacular discovery. Nevertheless, as a more technically focused NMR researcher, I wasn't so sure how to go about marketing my idea as a "biological system". I didn't have anything special up my sleeve; the ideas of Jean Jeener and Paul Lauterbur only actually existed in my head at the time. However, I decided that the upcoming convention in Kandersteg meant it was time to conduct my first experiments. On 24 May 1974, I wrote once again to Wes Anderson that "this week, we will conduct our first imaging experiments with the new Fourier technique". This marked the start of my foray into imaging methods. In theory, my PhD student Enrico Bartholdi and I already had everything well laid out; even the computer programs we would use were more or less written already and we just had to polish the edges. So my postdoc researcher got to work. We decided not to put a mouse in our magnetic field, instead opting for a simple water-filled test tube, which still let us reproduce the principle. This is how we created the first magnetic resonance image using the two-dimensional method and the Fourier transform.

While the results that I was able to present in Kandersteg were not perfect, with many attendees graciously judging them to be "premature", this did not stop me continuing down this path. I intended to patent this method. As I was still employed by Varian as a consultant at this time, it took the rights to the patent and in return paid me a patent fee of 200 US dollars, an amount I was very pleased with at the time. While many researchers of course went on to refine and improve upon the method, and apply it to many other fields, it is this two-dimensional Fourier transform method that still forms the beating heart of all MRI imaging processes in medicine. I then applied to the Swiss National Science Foundation for more funding to continue my work on imaging and if possible, to develop an MRI scanner myself. My application was turned down immediately. The reason given was that basic scientific research does not include optimizing the technology surrounding an already known basic principle. It is instead the job of private industry to finance any necessary optimizations to a given piece of technology.

We did not let this discourage us. But we focused on developing the two-dimensional method for using NMR to analyze chemical and biological molecules. We were once again completely carried away with our research. And as we confirmed later, even "premature" scientific babies can grow up to be real stars.

First international conference on MRI imaging in Nottingham, 1976. Richard Ernst is visible in the center left (with glasses), right in front of him Paul Lauterbur, behind him Sir Peter Mansfield (slightly obscured). Both later also received the Nobel Prize.

This actually marked the end of my own personal Dark Ages. The new developments opened up the narrow confines of our previously one-dimensional view and gave us a whole new understanding of "nature's spies", the magnetic hydrogen nuclei. Even today, I still feel great satisfaction at that feeling we had when we were able to broaden our view of the world like we did. Our group was imbued by a sense of euphoria, even though you wouldn't hear us letting off any champagne corks, mainly because the path we were taking was not easy. The pool of equipment I had at my disposal at ETH was still far from being state-of-the-art and the results of our experiments were often disappointing. It was a case of turning a blind eye to these shortcomings and trusting in your own idea more than the poor results. Many people would have given up. But we carried on experimenting, diligently performing our calculations and analyses. We drafted the designs for new pieces of equipment and developed the computer programs we needed for them. It later transpired, however, that it was possible there and then to expand the concept, using multiple pulse sequences to advance into any number of further dimensions. This gave rise to 3D, 4D, even up to 7D spectrums, which took our field into a completely new stratosphere.

Explosion in laboratory D2

Things weren't all plain sailing in the lab. One Thursday morning – it must have been sometime in 1976 – I was expecting another quiet day when I got off the train at Zurich main station and began to make my way up to ETH. But when I got into our building, there was chaos everywhere: There had been a large explosion. Unfortunately, the accident had happened in laboratory D2, one which was assigned to me. The rescue operation was already well underway, with no less than Professor Günthard taking personal charge. He evacuated everyone from D2, while I stood to the side, not knowing what to do. I was paralyzed by the shock of the events. By contrast, Günthard saw the situation as a chance to put his leadership qualities to the test and was enjoying doing so.

A technician from the mechanical workshop was worst affected; he had been anodizing a piece of aluminum in the corner of the lab when the explosion occurred. As was common practice at the time, he had been working under a fume-hood treating the light alloy with sulfuric acid and black mordant. The blast from the explosion at the other end of the lab was felt all the way over where the technician was working; the sulfuric acid and mordant splashed over his face and body. When I got there, he was sat desperately on a chair waiting for help. Shortly afterwards, he was led into an ambulance, which took him to the University Hospital Zurich, where he had to undergo numerous skin grafts. I visited him a number of times afterwards. Even though I could have done nothing to prevent it, I of course felt a great deal of sorrow about the accident, and I later paid him 1,000 francs of compensation from my own pocket. Happily he did not suffer any lasting damage.

Unfortunately, it transpired that one of my best PhD students had been responsible for the accident. He had been wanting to grow some crystals and had used about 100 grams of dimethylammonium perchlorate – the name alone sounds dangerous enough. He had dissolved this substance in a glass beaker containing a solvent and had planned to later allow the mixture to slowly crystallize on a hot plate. He then left the laboratory for a short time, and that is when it happened. The other technicians present that day later reported that the contents of the beaker had quickly turned brown, had started smoking like mad, and then exploded. My PhD student obviously hadn't realized that the reaction would get out of control quite to this extent, and so hadn't informed anybody else about his experiment before leaving the room.

The sulfuric acid also destroyed floors, tables and equipment in the lab. A number of windows were smashed, spreading shards of glass down onto the pavement outside on Universitätsstrasse, where thankfully nobody was walking at the time. Günthard was acting as if the entire building was on the verge of collapse and straightaway doused all the rooms in water to prevent any further damage. I thought this was an overreaction, but I didn't have the self-confidence to say anything. We did not speak again about who was responsible for the accident, but I suffered many a sleepless night and could no longer look him in the eye without a shiver running through my whole body. I later had to send a damage report to the university administration and the entire laboratory was renovated, with all furniture and fittings being replaced. In the end, we had one of the most modern labs in the whole building.

Unfortunately, the relationship between the responsible PhD student and Professor Günthard was already tense before the accident, because Günthard had originally wanted him to join his research group, but the outstanding student had instead opted for my group. I had a reputation for taking a more human and liberal approach to my students. I always made the effort to see the person behind the scientist. My aim was for my employees to have the feeling – the assurance even – that the success we enjoyed as a group was also down to their hard work. I would on occasions invite them for dinner in my home, or we would go on group hikes in the mountains, where I would enjoy leading the way. We treated each other with respect, and I would always refer to the PhD students using their last names. It was not until they had completed their dissertations that we moved onto first-name terms.

Richard Ernst in his office at ETH Zurich, at the age of about 44 years.

Being disciplined at work has always been of paramount importance to me. I considered "vacation" a dirty word, it simply wasn't in my vocabulary. Of course I had to grant my employees their vacation days from time to time, even if I would rather not have done. Personally, I never took any days off, at least not voluntarily. This did not mean, however, that my PhD students and postdocs always had to be sat there in the lab or the office. It was what they achieved that interested me, not the fact they were sat working where I could see them.

It is a self-fulfilling prophecy that excellent students tend to attract even better students who will follow in their footsteps. I was lucky enough to attract some of the best students for my group. Many of my PhD students and postdocs would later go on to be professors themselves, picking up many prizes and awards along the way. Particularly in my later years, I had a number of excellent female PhD students, despite this very technical field tending to attract primarily men – this was almost exclusively the case early on, and there is still a distinct gender split even today. If I were to try and list here all the wonderful staff I have worked with over the years, I fear that I would do someone a great disservice by forgetting them in my old age. Of course I also worked with employees who decided to pursue different paths, often of their own volition. It was sometimes the case that one of my PhD students just wasn't up to the job and I would take them aside and have a serious chat, perhaps even recommending that they pursue a career elsewhere. But they were all wonderful people, most of them true individualists. Many could also be difficult characters, but somehow we always managed to combine all these different personalities to make a successful group. At the end, we were so close that many people would refer to us simply as "Richard Ernst's NMR family".

Escape from the family crisis

There were dark clouds approaching in my own family life, however. The causes were rooted in my distant past and also the close relationship I had with my mother. It came as a great shock to her when Magdalena and I left Switzerland in 1963 to seek our fortune in California. While I never had any doubts about our move – and even my mother knew that I wanted to take this step – this came at a time when my sister Verena had got married and our youngest sibling, Lisabet, had moved out. Suddenly my mother was completely alone in a large house on Gottfried-Keller-Strasse in Winterthur. It was therefore only logical that after we returned from the US, now as a family with two daughters, we would move back into our family home, especially given the apartment on the first floor was now free. This had a number of advantages at first: We didn't have to spend a long time searching for an apartment, my mother loved having us and the kids around, the children had plenty of space to play in the house and the garden, and the location near to Winterthur station was perfect; I could easily commute to my lab in Zurich every day by train.

Upon returning to Switzerland, we experienced a second culture shock. Compared to the US, life in Switzerland was far more old-fashioned, somehow restrictive even. In Switzerland, women did not even have the vote. While the research environment was very competitive in the US, we always treated each other fairly and cooperated well. By contrast, ETH Zurich – the "poly" – was still imprisoned within the confines of a hierarchical system dating back to the 19th century. However, it was the small differences that we really noticed: The narrower streets instead of the wide highways; cramped little grocery stores instead of massive shopping malls; many tasks still performed manually instead of using domestic appliances.

On 9 December 1972, our third child, Hans Martin, was born. "Finally, a son," thought many people, especially representatives from the wider Ernst family. The christening was a big party: The grandmothers brought gifts; aunts and uncles stopped by to greet us; and everyone was delighted that we finally had a child to carry on the Ernst family name. Only in retrospect do I realize that as a father, I had fallen into the very sort of patriarchal role from which I had suffered so greatly as a child. At the time, Hans Martin's birth came as a shock to his eight-year-old sister Anna in particular. Previously, she had been my mother's favorite child and had enjoyed a great deal of affection. Suddenly she was forced to take a back seat and reacted fiercely. Magdalena told me that from this day on, Anna never wanted to wear a skirt, as girls did in those days, but instead would only put on trousers. She became a real tomboy, much preferring to play football outside than stay inside and play with dolls.

It's true to say that I only heard about a lot of the things that happened in my family after the event, second-hand from Magdalena. As I had told her before we married, I had little time for my family. I was happy if I could sit there working in silence, sometimes pausing to carry out repairs that needed doing around the house. I liked to do so, often also for my mother. This meant that I had nothing to do with my children's upbringing or their schooling; I didn't even know where they went to school. I would often skip the Sunday walk with the children, my wife looking on wistfully at other young fathers pushing their strollers through the park. When we went on vacation – something I only ever did very much against my will – I would always take my work with me. Often when the weather was nice, we would go for a walk together as a family, but if the opportunity presented itself, I would sit in the bath or hide away in a bedroom of our holiday house, which I would have converted into a temporary office, immersing myself in two-dimensional spectrums or scientific literature.

I spent most of my time on science and research. I would leave the house early in the morning and come back late at night. Even on weekends, I would much rather spend my time thinking about magnetic atoms or pulsed radio waves than have to spend time with my son or play with my daughters.

While my workplace in California was not far from our home and I could therefore play a more active role in our family life even though I worked very long hours at the time, the commute between Winterthur and Zurich just seemed far longer by comparison. This was the case not only because of the distance, but also due to how I felt at the time. I still suffered a lot of self-doubt. I had convinced myself that I was still yet to achieve anything in my scientific career. Despite my success, I was very insecure, and perhaps had retreated too far within myself. It was the problems with the equipment and the continual search for scientific breakthroughs that were causing me sleepless nights, not the children, not the concerns of my wife. I was literally married to my equipment.

When you face so many challenges in your job, it's easy to ignore any problems at home, or you can just assume that everything will sort itself out. But gradually the cracks were becoming wider and the problems were stacking up. The relationship between Magdalena and her mother-in-law was becoming ever more strained. Since my father's death, my mother had been the undisputed head of the house, and she had made it clear to Magdalena how she thought she should run a household. If my mother thought Magdalena should be doing more in the garden, she would give her a rake or some pruning shears for her birthday. She would also interfere in how our children were being brought up – in her eyes, Magdalena couldn't do anything right. Once when my wife objected to this, my mother told us in no uncertain terms that it was time for us to move out and didn't speak to us for a week. Afterwards, she was sorry, but Magdalena had had enough.

I somehow found myself caught between the two of them. For my mother, I was the center of her life, especially since the death of my father. But it was also important for her that I never abandon her. However, I was rarely at home; the longer you spend working in science, the more often you find yourself traveling, whether to conventions, symposiums, summer schools, lab visits – I could find myself anywhere in the world. This is all part and parcel of a scientist's life, something I knew and accepted right from the start. It was not easy for Magdalena to be on her own with the children and her demanding mother-in-law. She slowly sank into a depressive exhaustion without me really noticing. At this point – the situation reached a head in early 1976 – our marriage was also on a knife edge and the children were also suffering from the situation. Our oldest daughter Anna was having obvious problems at school and it was not a certainty that she would be able to go to secondary school. Luckily, she made it in, and later on she even managed to transfer to an upper secondary school. At this time, a kind, understanding social worker who worked for the church helped Magdalena overcome her depression. She rescued our family. She and others advised us to look for help for Anna too. The idea was that she attend group psychotherapy, but then a youth psychiatrist recommended to my wife that we look further afield. If we only treat Anna now, explained the specialist, the next child, the weakest, will then only have other problems. He convinced Magdalena that instead, the whole family should attend family therapy, of course under the condition that I, too, agreed to go.

Convincing me was no easy task for Magdalena. My attitude towards psychological and psychiatric therapy had not changed since my nervous breakdown, not least because at the time I had managed to sort things out myself – or so I thought. I was extremely resistant to advice. I was a proud man, perhaps too proud, but Magdalena and the children were suffering so much from the whole situation, and they implored me to take part in the therapy sessions. Thankfully I finally agreed. For six months, we regularly visited two specialists from the youth psychiatry service. During the sessions, we learned mainly how to communicate with each other. Every day, we would take an evening walk in our neighborhood and talk about what we had done with our days. And a miracle occurred: The therapy helped us out of our mess.

Anna started doing better at school. Six months later she passed the entrance exam for an upper secondary school, after which she started to build an amazing, successful career. First, she trained as and then subsequently became an excellent kindergarten teacher. Because she would always play football with the children, she was popular with the boys in particular. She then attended arts college, became a handicrafts teacher and went on to hold a number of successful exhibitions as a creative artist. From Austria, where she is married, she visited a college for art therapy in Munich. Most recently, she trained as a psychotherapist at the University of Vienna, and today she runs a successful practice in the city. Katharina and Hans Martin also went their own ways. Katharina became a primary school teacher and then a speech therapist. She finished top of her class in her high-school-leaving exam. I had plans for my son Hans Martin, and would have liked for him to become a scientist. Even at an early age he had an encyclopedic memory and developed an unbelievable passion for railways. He knew everything about all the locomotives running on the Swiss network – models, numbers, coats of arms – and whenever I came home from work, he would always ask what sort of train I had traveled on. Everything seemed to be going nicely when he achieved the top grade of 6 for mathematics in his high school entrance exam, with his teacher saying that he had never had a pupil as good at mathematics as my son. But my dreams of one of my children following in my footsteps would go unfulfilled. Hans Martin wanted to study psychology. After getting his degree, he first worked in market research and finance, until he decided to train as a therapist in his 40s. Today, he works as a psychotherapist in a prestigious private clinic.

My children have all grown up to be very independent. They had no other choice as I was not a good father to them. When journalists would later ask me about my private life, I would say that my most important contribution to their upbringing was acting as an example of how not to behave. Science and research were always my top priorities. I was always sure that as a researcher, you had to have an attitude which I would like to describe as the courage to exercise self-contempt. Making myself happy simply wasn't part of my character. Unfortunately I don't have any grandchildren, which – as the son and heir of the Ernst family name – saddens me, and this is not likely to change during my lifetime.

The Ernst family in their newly built home in Winterthur in 1978: Anna, Richard, Magdalena, Hans-Martin and Katharina Ernst (from left to right).

Previously, my family heritage and the way I was brought up to behave would have prevented me from speaking so openly about these problems, but now, at the end of a long life, I don't have any trouble talking about them. And – at least according to Magdalena – I really benefited from the therapy sessions we attended. "He can even laugh now," old friends would say about my transformation from a serious man, full of self-doubt, to an almost humorous person, although it goes without saying that I haven't rid myself completely of my doubts and self-critical attitude.

Even before we attended the family therapy, we had decided to build our own house. We no longer lived at number 67, Gottfried-Keller-Strasse. Even though I often think wistfully about this beautiful late-nineteenth-century villa, with its high ceilings and stylish stairwell, it was good that we now had our own home.

As my career developed, I did not have more time for my family either – quite the opposite, in fact: The further I progressed, the more responsibilities I had, and these responsibilities frequently involved me being away from my family. Thankfully Magdalena was able to deal well with this hardship. While she gave up her job as a teacher to move to the US with me, she hadn't neglected her own interests. She was not one of those women who sacrificed her own personal development for her husband and children – and this is something that came as a real relief to me most of the time. She would always find time to do what she was passionate about, whether playing the violin, singing in her choir, listening to chamber music or enjoying art and literature, and the Tibetan scrolls – the thangkas – were a shared interest for us both.

The road to success

With hindsight, it is hard to say exactly how much the success of our family therapy sessions helped get my scientific career back on track, but I can definitely say it had a positive influence. Indeed, my career really began to take off from the mid-1970s. Suddenly there was a lot of interest in our methods among experts. It was at this time I finally gave up my role as a consultant at Varian. I had received a query from the company Bruker, which had had a close relationship with ETH for a number of years, in particular with the Laboratory of Physical Chemistry. Hans Heinrich Günthard and Hans Primas, my two doctoral supervisors, had previously worked together with Bruker's predecessor Trüb, Täuber & Co. AG to develop an NMR spectrometer, before the company was bought by Bruker and renamed "Spectrospin". Later on starting in 1969, Bruker-Spectrospin went on to become the first commercial company to adopt the Fourier transform method I had developed at Varian. At the time, the driving force behind Bruker-Spectrospin was a man named Tony Keller, with whom I would later closely collaborate. He was a visionary manager, but also knew enough about science to be able to correctly judge the opportunities and risks presented by my method. While the patent still belonged to Varian and was registered under Wes' and my name, Bruker was able to make use of it thanks to a licensing agreement. As was usual for this sort of patent, Varian would profit most from its success. As I had technically registered the patent as an employee, all I had was the 100 dollar patent fee.

It was back in 1969 that I initially wrote to Wes Anderson, who at the time was still head of Varian's science department, to let him know about Bruker's keen interest in my work. Out of loyalty, I remained faithful to Varian for many years, even though the company's management clearly underestimated the opportunities presented by NMR technology. But then the right moment came: When Günther Laukien, the founder and guiding spirit of Bruker, approached me in the mid-1970s to see if I wished to work together with him, it was an easy decision to make. By then, Bruker had become the market leader in NMR equipment. Firstly, this was thanks to the consistent implementation of Fourier transform technology in chemical analyses, but secondly also a result of the development of NMR spectrometers with superconducting magnets from 1970 onwards. By using superconducting magnets, which allow electricity to flow through the magnetic coils with almost no resistance, it is possible to create far stronger and more stable magnetic fields than when using traditional electromagnets. Stronger magnetic fields make it possible to measure and discover resonance effects that were not previously possible.

Bruker-Spectrospin's and my close collaboration was primarily in the area of multi-dimensional NMR spectroscopy for biomolecules. In 1976, I joined forces with my ETH colleague Kurt Wüthrich. and we managed to get funding for a joint project on the "development and application of 2D NMR for proteins". Together we laid the foundations for an extremely successful research community, which was known among experts as the "Zurich Group". It was the beginning of a partnership that lasted more than a decade, before finally coming to an end in 1986. It produced many joint publications, but also numerous patents, which Bruker-Spectrospin was able to make commercial use of. Kurt Wüthrich and I had previously had casual contact, usually when it came to technical questions or getting access to the equipment I was responsible for. He returned from the US in 1969. His earliest work at the University of California, Berkeley and at Bell Labs focused on researching on large and complex biomolecules such as blood proteins using the

NMR method. By contrast, I was the engineer, who was aiming to build the most perfect measurement devices possible. One day – it would have been in 1975 or 1976 – Kurt Wüthrich paid us another visit in our lab. On the table was a spectrum which we had produced using our innovative two-dimensional model. It came to him like a bolt of lightning. He sat down, his head in his hands, and retreated into deep thought. He saw straightaway just how useful this method could be for his work in examining highly complex biomolecules.

We were mainly interested in proteins, the fundamental components that do just about everything in the human body: As enzymes, they act as production machinery, gate-keepers in cells, regulators for genetic functions or even just structural elements. A protein is similar to a colorful, rolled up pearl necklace. Using this metaphor, we can say that the colors of the pearls could correspond to the basic building blocks of the protein, the twenty amino acids. The way the necklace is rolled up has a major influence on whether and how the proteins function. A pearl necklace can contain dozens or even hundreds of amino acids, all of which also contain hydrogen atoms, which in principle can be examined and located using NMR technology. We were fascinated by the prospect of being able to use "our" method of having hydrogen atoms act as spies to decode the exact structure of these pearl necklaces – not only the sequence of the pearls, but also their shape and the way they are folded. NMR spectroscopy also had the decisive advantage that it was possible to study the proteins as they occur in nature, because with NMR experiments they are measured in an aqueous solution, where they would be folded as they are in the human body.

We set to work immediately, each assisted by two or three postdocs. I provided the methodical and technical expertise, while Kurt Wüthrich carried out experiments on the proteins and researched specific applications. We spent more than a decade advancing the theoretical and practical development of multi-dimensional NMR spectroscopy. But the longer our partnership lasted, the more difficult it became. He formulated objectives and then at the end, could provide real results when he once again managed to decrypt the structure of a protein. By contrast, my work was more theoretical and harder to explain, and therefore received very little media attention. You could say that I had built the Ferrari, but it was Wüthrich who won the race in it. We would frequently argue about how our joint contributions should be cited in publications. Given our close partnership, it was inevitable that we would end up citing each other's work. I would refer to his results to show that my methods worked, while he had to cite my methods which he was using to get his results and decipher the structure of the proteins. This meant that our achievements were inextricably linked. While we both carried out a lot of research outside of our joint project, I increasingly sensed that the balance needed for a productive partnership was being thrown out of sync. Our personalities were too different. Kurt Wüthrich was the self-confident high flyer, who would always say how he thought things needed to be done. I remained the somewhat shy and introverted scientist, who would much rather hide away in his silent lab, looking for his next big breakthrough. We ended our joint project in 1986. We each went our own way and from then on, our contact was limited to a professional minimum.

In scientific terms, our project was a great success. There is no doubt that in his experiments, Kurt Wüthrich became the first person to consistently apply the two- and three-dimensional NMR methods to biomolecules and decode the structures of ever larger and more complex protein molecules. Technically, this was a huge challenge, which required him to develop special pulse sequences. Even analyzing the data and then using it to create meaningful spectrums was a stroke of genius. By the end of the 1980s, the potential offered by the method was obvious. Nowadays, nuclear magnetic resonance has surpassed other processes such as X-ray crystallography for many applications. At times it is even possible to dynamically track the "life" of proteins; we can see them "in action", so to speak.

The rapid development of NMR processes was a key part of the drive to move into new technologies in science and medicine in the second half of the twentieth century. While this progress was based on findings from basic scientific research, it would not have been possible without the involvement of private companies. At first, the work Kurt Wüthrich and I carried out was financed by

ETH, but the university soon withdrew this funding to saves on costs. It was then that Bruker-Spectrospin stepped in and expanded its commitment, and a Swiss funding organization specializing in application-based research, the Commission for the Advancement of Scientific Research, also stepped into the breach. Without this belief in the potential of the method, it would not have been possible for NMR technology to progress from being an expensive and rather inefficient measurement method at the start of the 1970s to the globally ubiquitous and highly efficient technique that we see today, used widely in chemistry, biology and medicine. And the best thing about it is that Switzerland always played a major role in this story.

It was in 1983 that my work finally started receiving the recognition it deserved. I received the Gold Medal Award from the renowned Society of Magnetic Resonance in Medicine, which was presented in San Francisco. This was followed in 1986 by the Marcel Benoist Prize, one of the most prestigious science awards in Switzerland. At the time, I thought my life had peaked – well, almost peaked.

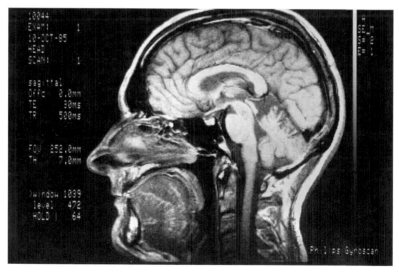

The MRI image taken on October 10, 1985, in Zurich shows a cross-section of Richard Ernst's head.

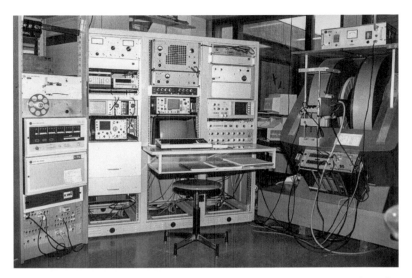

An NMR spectrometer at Richard Ernst's workplace at the ETH Zurich in the early 1980s. The circular magnets that produce the magnetic field for the NMR experiment are clearly visible.

The Light at the End of the Tunnel:
The Nobel Prize, 1991

The bet

In autumn 1989, my colleague Willi Simon and I were joking around and he bet me 500 francs that I would win the Nobel Prize that year. I was sure I wouldn't, and of course went on to win the bet. I was happy to receive my winnings and spent them on an expensive book about Tibetan medicine, which fed my second passion besides physical chemistry: Tibetan art and philosophy. Sadly Willi Simon died shortly afterwards, meaning he never got to see that his prediction came true, just two years later.

The idea of me winning the most important chemistry prize there is – the fabled Nobel Prize – was something that I could not have imagined in the years leading up to it, even if it was something I secretly dreamed about. My gut feeling was that it was beyond me, and I believed that all the facts would back up this assumption. Was it not the case, after all, that chemistry journals in the 1960s had rejected my first work about the Fourier transform? Did my first two-dimensional spectrum, which I presented to the legendary scientific conference in Kandersteg in autumn 1974, not get greeted with a high degree of skepticism? Don't Nobel Prizes only get awarded to people when they are old and gray? After all, I had only turned 58 that year. Doubts and a crippling lack of self-confidence followed me around constantly, even during the most successful years of my scientific career in the 1980s.

In the meantime, I had received a number of scientific prizes: After the 1986 Marcel Benoist Prize, I was awarded the 1989 Kirkwood Medal by Yale University, and then two years later, the important Wolf Prize in Israel, and the Louisa Gross Horwitz Prize from Columbia University in New York. While I was, of course, pleased to receive all of these awards, it did also mean that I had to deliver more lectures, that I was invited to more conferences, that I had to attend more meetings. I was always on the road. I even heard rumors that in recent years, I had been nominated on several occasions for the Nobel Prize, even though this was supposed to remain secret. There was a lot of chatter among the scientific community; researchers do like to gossip – after all, they are only human.

And then on 16 October 1991, I actually received the call from Stockholm while I was sat on a flight from Moscow to New York.

Journalists asked me over the aircraft radio about the significance of this prize, and I was quick to assure them that the award represented a major success for Switzerland, for ETH Zurich, and for my research group. It goes without saying, however, that the prize was also an unbelievable honor for me. I had sacrificed so much for so many years; day after day, I would commute from Winterthur to Zurich, at my desk for seven o'clock in the morning, not returning home until eight o'clock at night, just to work for another three hours before bed, and I had barely taken any vacation. Of course, I had achieved a lot, made a great many friends, got to know numerous brave, outstanding colleagues. But I had also had to put up with resentment, jealousy, and simple ignorance. My boss at ETH had always treated me like little "Richi", even when I was already a professor, and instead of acknowledging my achievements, he went out of his way to disparage them. "That's nothing new," he would often say when I came to him with an idea, or: "Well that's obvious!" And then I received the Nobel Prize. It was more than I could have dreamed of. I had a tear in my eye when I read the reason for me being awarded the prize in the press release: "for his ground-breaking contributions to the development of..." I slowly began to believe that it was true. "There can be no doubt as to what Richard Ernst has achieved," wrote the *Frankfurter Allgemeine Zeitung*.

Because I was sat on a plane to New York when the prize was announced, my wife and son Hans Martin were called to ETH that day, where a press conference and a small celebration in my honor were due to be held. I read a report later on in the newspaper that people were even cheering me! My son told those assembled that he wasn't surprised when he heard the news. "This morning I just felt there was something special in the air. When the phone rang, I somehow knew already what we were about to be told."

Then it was my 82-year-old mother's turn to speak. She told reporters that she was trembling when Magdalena gave her the news, and that all she could do was head into the garden. "Getting fresh air always does me good, even when I am in a real fluster because I'm happy!" She said she was amazed how kind and friendly everyone was to her now that she was suddenly so famous. "Now I'm the most important woman in Winterthur," she smiled.

Later on, I took a secret pleasure in reading all the reports. My former high-school biology teacher told one newspaper that "Richard Ernst was an outstanding student across the board. He and the whole class were real characters. I have very fond memories of them because even though they posed challenges, they in turn rose to them as well." He went on to say that I was a student he always knew would tread his own path. I had to smile when I read this, thinking back to the reprimands I received and my poor grades in French and Latin.

I was particularly delighted with my reception in my home town at the Stadthaus Winterthur. Who wouldn't want to at least once feel like the protagonist in Dürrenmatt's drama "The Visit", returning home from far afield to fulfill his hometown's every conceivable wish. At the reception, the then mayor Martin Haas told those assembled that Winterthur had never previously celebrated such an important event. "He is a real Winterthurer!" The local newspaper, the *Winterthurer AZ*, featured a particularly flattering headline: "Winterthur basks in the fame of its Nobel Prize winner." This also reflected on our family, the Ernst family.

Of course, this soon all got a bit too much for me, so I made every effort to come across as normal. I found myself saying things like "I'm just a normal guy from Winterthur" or "please, call me Richard, not Professor". I tried to almost pretend that the prize wasn't all that important, that the aim of my work was never to win awards, that my main goal has always been research – or educating the next generation.

Roller coaster of emotions

Winning the Nobel Prize brings with it a roller coaster of emotions: You are overjoyed to receive the award, but then highly embarrassed in the same breath. Of course you are pleased, but you straightaway think about your many colleagues and employees with whom you have worked for many years. And it drags you straightaway into the limelight. Fundamentally, however, I had never truly been able to overcome my shyness in the face of others. I still hated being the center of attention, and could think of nothing worse than having to make small talk and chat idly with strangers. I much preferred to work on my own in my lab, where I could do what I wanted.

And for me, the situation was even worse. It is tradition for a Nobel Prize in each category to be awarded to up to three laureates, particularly because in modern scientific research, success comes not due to the work of one lone wolf, but as a result of – in many cases, interdisciplinary – teamwork. This meant that my being announced the sole winner of the 1991 Nobel Prize in Chemistry was an exception, rather than the rule, which I found particularly embarrassing.

The main reason for this was my colleague at ETH, Kurt Wüthrich. As I mentioned before, during our ten-year partnership, he played a pioneering role in advancing NMR methods for large, complex biomolecules. I had contributed the experimental and methodical principles underlying two-dimensional NMR spectroscopy, while he expanded the application of these methods and made a number of groundbreaking discoveries. As a sports teacher by training and a passionate footballer, he had relentlessly pursued his scientific goals with an unparalleled level of commitment. He had also been talked about as a potential candidate for a Nobel Prize on various occasions. The fact that it was only me, an introverted, self-doubting scientist, to receive the award was a tough blow for him to take. He suddenly saw his hopes being swept away, because when the Nobel Committee recognize a certain field – in my case NMR – this often means that this field will no longer, or certainly not for many years after, be considered again.

High above the Atlantic, when I received the call from Stockholm, I realized all of this straightaway. The fact that I was on my way to meet Kurt Wüthrich in New York, where we were to be *joint* recipients of the Louisa Gross Horwitz Prize, made me extremely anxious. It was a disaster. Shortly after our meeting, he wrote me a letter in which he strongly requested that I no longer mention his results in my lectures. The fact that only my work was recognized also seemed unfair to me, but it had enraged my colleague.

From the very day I was awarded the Nobel Prize, our relationship deteriorated. The little contact we had became even less, and the idea of a joint project was simply inconceivable. Kurt Wüthrich buried himself even further into his research on biomolecules. It was not until some years later, when I was in the process of becoming an emeritus professor and there were discussions as to who my successor would be for NMR research at ETH, that we sat down again together. "Restaurant Neue Waid, dinner with Wüthrich," I noted down in my diary for 29 July 1996, "very relaxed, open conversation, as if nothing had happened between us. A very pleasant evening."

And then in autumn 2002, a miracle. On Wednesday, 9 October, I was on my way to a scientific conference in Munich when I heard on the radio that the Nobel Prize winner in Chemistry had been announced. And then I heard some news that delighted me. I noted it specifically in my diary: "11.45, NP chemistry announcement: 1 American + 1 Japanese for mass spectroscopy ... And Kurt Wüthrich!!!! A miracle! He doesn't have e-mail." I wanted to congratulate him straightaway, but couldn't get in touch. Instead a journalist from a Sunday newspaper called me. He probably remembers that I had received the Nobel Prize in the same field and asked me for some words of praise for Kurt Wüthrich. "I'll send him some," I wrote simply in my diary. I agreed immediately, even though I was in Munich and would only be able to write something worthy of his achievement by working into the night. I delivered the text on time on Saturday morning and wrote in my diary: "Now I feel somewhat more liberated."

The following Monday, I finally managed to contact Kurt Wüthrich by e-mail and he called me straight back. My next diary entry highlights the sense of relief this gave me: "Wüthrich is really delighted and can't think of anything better than us both having received one prize each. It's nice that we have made up our differences." I suddenly realized that we had gone our own ways and had both mutually accepted each other's decisions. A month later, he really surprised me by paying me an unexpected visit in my lab. "We enjoyed a very open conversation," I noted after the event. We were able to heal the rift that had emerged between us since 1991: "He apologized for his displeasure in 1991, but he felt it was the end for him in science, there was no resentment towards me (?). He didn't believe he would win an NP in chemistry, more likely in medicine with Lauterbur." We then talked shop a little about the other Nobel Prize winners and our field, nuclear magnetic resonance. I was really pleased to note that "The Zurich Group" was still alive and kicking.

The secret of the Nobel Prize

Outsiders might wonder why the Nobel Prize in particular is so coveted by scientists and how this award manages to extend its magical aura beyond the scientific community into the general public – a prize awarded in a country such as Sweden, which has indeed produced some excellent scientists, but in terms of research, does not come close to countries such as the US, the UK, or even Germany.

The history of the prize is intertwined with the fascinating figure of its founder, Alfred Nobel, the inventor of dynamite. Nobel's life was full of contradictions and made him the philanthropist he was, whose legacy even today has lost none of its sheen. To understand this, I would like to briefly tell his life story. Alfred Nobel lived in the 19th century and was an extremely wealthy, but lonely person. He was born in 1833 into a small industrialist family in Stockholm as one of eight siblings, and his father, Immanuel Nobel, was also an inventor. However, Immanuel soon went bankrupt and fled to St. Petersburg to escape his debts. He left behind his wife, Andriette, and their children, who struggled for survival, with the children having to sell matchsticks on the street. This period of extreme poverty had a lasting impact on Alfred. In St. Petersburg, however, the family's fortunes changed. Alfred's father was working in the munitions industry and had built up a tidy fortune, which enabled him to bring his family over five years later and provide the children with a good education. He sent his son Alfred to Paris to become a chemist. Here, Alfred Nobel met the Italian Ascanio Sobrero, who had invented a substance called nitroglycerin by combining glycerin, nitrous acid and sulfuric acid, but did not know what he should do with his invention.

Alfred, however, realized how this substance could be used, and back in Stockholm, he built an explosives factors to manufacture the substance. Unfortunately, Alfred's younger brother Emil was killed aged just 19 in an explosion at the factory. This very much affected Alfred, but also woke his inventive spirit. He developed and patented a detonator and invented a form in which the liquid nitroglycerin could be used safely: Dynamite. The business flourished and Alfred Nobel opened factories across the world, in Europe, the US, and in

South America. As his wealth continued to grow, so did his guilty conscience. Although he had invented dynamite for civilian uses, such as in construction, it was soon also used in wars.

Despite his wealth, Alfred Nobel was torn on the inside, melancholic and frequently depressed. Once, when he was 54, a journalist asked him to describe his life, to which he gave a somewhat sarcastic reply: "Greatest merits: Keeping his nails clean and never being a burden to anyone. Greatest defect: Lack of family, a happy disposition and a good stomach. Greatest and only request: Not to be buried alive. Greatest sin: Not worshiping Mammon. Important events in his life: None."

Alfred Nobel died of a heart attack on 10 December 1896 aged just 63 at his home in Sanremo, Italy. He had been ill throughout his life and did not have any children. He had therefore begun to think about his legacy at a relatively early age, ultimately deciding to make sure his estate was used for a good purpose. In his last will and testament, he instructed that his "capital, converted to safe securities by [his] executors, is to constitute a fund, the interest on which is to be distributed annually as prizes to those who, during the preceding year, have conferred the greatest benefit to humankind. The interest is to be divided into five equal parts and distributed as follows: One part to the person who made the most important discovery or invention in the field of physics; one part to the person who made the most important chemical discovery or improvement; one part to the person who made the most important discovery within the domain of physiology or medicine". As Alfred Nobel was also fond of poetry and was a philanthropist, he also founded the Nobel Prize for Peace and the Nobel Prize for Literature. It was another sixty years before the Academy added the Nobel Prize for Economics in recognition of the vital role that economics plays in the fate of the global community. It is noteworthy that in his will, the chemist Alfred Nobel wanted to recognize discoveries and improvements in the field of chemistry, as opposed to physics, where he referred to discoveries and inventions, or medicine, where he focused only on discoveries.

Alfred Nobel left most of his wealth to this foundation – some 33 million Swedish krona, which today is equivalent to around 330 million US dollars. His relatives and his most loyal staff – indeed even his gardener – received several hundred thousand krona or a lifelong pension. His immediate family were massively disappointed that they had only been left a few crumbs. They furiously contested the will and took all possible legal action to prevent his last will and testament being executed. They were ultimately unsuccessful, however, meaning that King Oscar II was able to definitively set up the Nobel Foundation in 1900, with the first Nobel Prize being awarded in 1901. Since then, Alfred Nobel's name has been synonymous with scientific and cultural advances.

The lost medal

There is an endless list of science prizes that have since tried to replicate the Nobel Prize, but none have come close. There are various reasons why the Nobel Prize has become so famous and has such an aura surrounding it. The list of those who have received Nobel Prizes over the years certainly helps and goes to show that the Nobel Committee is doing good work. While it is true that many people who deserved a Nobel Prize did not receive one, most recipients have certainly deserved the honor.

And of course there are many other outstanding scientists who are yet to be awarded a Nobel Prize. Many were just unlucky that they died too early. Other scientists are simply ineligible because there is not a Nobel Prize in their specialist area. Have you ever heard of a Nobel Prize for Mathematics, a Nobel Prize for Environmental Sciences, or a Nobel Prize for Process Engineering? Other people were simply overlooked, so nobody should feel bad about not receiving one.

Another factor behind the Nobel Prize's continual growth in popularity is the significant monetary award that accompanies it. Ever since the prize was awarded for the first time, this money has always meant a great deal to the winners, many of whom already earn healthy amounts from their scientific work. Even in 1901, the 150,000 Swedish krona paid to winners was around twenty times the annual salary of a professor. When I received my Nobel Prize, this amount had risen to the equivalent of 1.5 million US dollars. Of course, I didn't come across a journalist who failed to ask me what I would do with the prize money. The answer wasn't simple. Everyone expected me to say I would reinvest the money in research. I did indeed do so, but not in scientific research – instead I used it to further my knowledge of my second great passion: Tibetan art. That, however, is another story, and one which I will talk about in the next chapter.

Each Nobel Prize winner receives a diploma in Swedish.

It is not only the prize money that has become a thing of legend: Winners also receive a diploma and a medal made of pure gold, with the winner's name etched into it. Even more legendary, however, are the stories of the various medals that have been lost, sold, auctioned or even melted down. James Watson, who discovered the structure of DNA and won a Nobel Prize in 1962, is said to have auctioned off his medal for almost five million dollars. During the Second World War, physicist Niels Bohr dissolved two medals in acid, which he was hiding from the Nazis on behalf of two German-Jewish winners in his lab in Copenhagen. After the war, the gold was then extracted from the solution and the medals were re-cast.

I kept my medal in a case in my office at home and didn't really think about it until I was asked to show it off at ETH. I quickly realized that it was not where I thought I had left it and a desperate search through the house ensued – but with no success. I even suspected that someone could have stolen it. The following night, however, I had a dream in which I could see the medal in its case in a specific place on one of my bookshelves. On waking up the next morning, I rushed over to the shelf and found it lying there, exactly where I had dreamed. I quickly decided to donate it to the ETH Archive, where it is still safely stored away today.

Magdalena meets King Carl Gustaf

The medals and diplomas are presented to winners on 10 December each year, the anniversary of Alfred Nobel's death. The ceremony and the subsequent Nobel Banquet are unrivaled in their pomp and glamor. The traditional celebrations are attended by Sweden's Royal Highnesses and feature many regal conventions that you would not expect to see in a social democracy such as Sweden. For we Swiss, who normally only come across kings and queens when playing cards, this was somewhat out of the ordinary. The dress code is very strict – men have to wear full evening dress and while things are somewhat more relaxed for women, it is an unwritten rule that they must wear a long dress. This did not go down particularly well with my daughter Anna, who only agreed to dress accordingly after repeated requests. I didn't have a dress suit hanging in my wardrobe at home either. Luckily, though, there were a number of enterprising shops in Stockholm, which were more than happy to rent out the requisite outfit for around 250 dollars, an amount that didn't eat too heavily into my prize money. On the morning of the ceremony, the laureates are invited to a final rehearsal – always chauffeured by their own driver – where they have to practice every movement and every step for the prize ceremony later in the day. This begins from the moment you are allowed to leave your designated seat and lasts until you are presented the medal and diploma by King Carl Gustaf, before you bow to the queen and the royal family as well as the audience in attendance.

Trying on the obligatory tailcoat for the Nobel Prize ceremony. Master tailor Jarl Dahlquist is making the final adjustments on the unfamiliar garment.

A happy day: on the road in the car specially provided for the award winners to one of the many ceremonial events surrounding the award ceremony.

Ready for the award ceremony: Magdalena, Katharina, Hans-Martin and Anna (from left to right) in the Globe Arena in Stockholm.

Each prize winner can be accompanied to the ceremony and the subsequent royal banquet by their closest family and other people important to them, up to a total of twelve people. I was happy that all my children came with us. My mother also attended; her trip to Stockholm was a highlight in her life. I realized that I had forgotten to invite Wes Anderson, my former boss at Varian, who had helped me so much in my career and had remained a good friend. Over the years, however, we had lost touch somewhat, mainly because we had both been so busy working in our respective fields. I quickly remedied this, getting a ticket for both him and his wife Jeannette, and I was extremely pleased that they were both able to attend. Wes took the opportunity to once again show his generosity, warmth and spontaneity: He recorded the ceremony and banquet on his video camera and later on, gave me the recording as a personal memento.

Queen Silvia of Sweden congratulates personally on the Nobel Prize. In the background, King Carl Gustaf.

Carl Gustaf, King of Sweden, presents Richard Ernst with the Nobel Prize medal on December 10, 1991. December 10 is the anniversary of the death of Nobel Prize founder Alfred Nobel.

The traditional banquet will be held in the Blue Hall of the City Hall of Stockholm. In the center of the picture on the right Magdalena Ernst between King Carl Gustaf and Erwin Neher, Nobel Prize winner for Medicine.

The royal banquet is attended by over 1,000 guests and takes place in the medieval Blue Hall at Stockholm City Hall. The guests of honor – the royal family, the laureates with their blue-blooded dinner partners, their relatives and certain politicians – enter the hall at a measured step via a grand stone staircase, accompanied by solemn organ music and a fanfare. These ceremonial traditions alone take a good amount of time. The flowers on the table are imported directly from Italy, freshly cut in the gardens of Alfred Nobel's old estate in Sanremo. The seating plan follows an old tradition, and etiquette dictates that you should pay more attention to your neighbor on your right than on your left. I was unlucky in this regard and found myself sitting between the elderly wife of a diplomat and the wife of a Swedish field marshal. It goes without saying that I would love to have been sat next to Queen Silvia, who was wearing a stunning red dress and her crown adorned with many sparkling diamonds. But at least Magdalena was in luck. During the banquet, she sat with King Carl Gustaf directly on her right and the German winner of the Nobel Prize for Medicine, Erwin Neher, on her left. Afterwards she told me how marvelous it had been to talk to the King over dinner, mainly about traveling to distant countries.

The celebration in 1991 was a jubilee event and as such, was especially ostentatious. All living laureates from previous years had been invited to Stockholm, and the banquet menu featured four courses, rather than the standard three: Nettle soup with quail eggs, marinated salmon tartar in red pepper sauce, roasted duck breast in a special sauce as the main course, and a dessert to finish. Each course was served with elaborate, almost exaggerated choreography, announced against the backdrop of another fanfare, and then brought to our tables by a myriad of liveried waiters and waitresses.

After dessert – in 1991, it was a Nobel gala-inspired ice cream parfait on a silver platter – all the laureates would give a three-minute acceptance speech to the Royal Society. The first to speak was South African Nadine Gordimer, who had won the Nobel Prize for Literature, and she enchanted the audience with a charming, witty speech. I didn't even try to emulate her, but I couldn't resist a few critical words about the pomp of the whole occasion. I'd like to include my speech below, as it gives an idea of just what I was feeling at the time:

"It is indeed a great moment for me to stand where I am standing to express my deep gratitude to the Nobel Foundation for this extraordinary honor. Obviously, most of the glory should fall on those standing behind me, my teachers, my colleagues, my coworkers, my school, my 700 years old country, those whom I represent here as their scientific spokesman. The presence of all the former Nobel laureates gives me a feeling of being carried by a swarm of wild geese, some real high fliers, like in Nils Holgersson, and I am afraid of falling down.

Science prizes have a tendency to distort science history. Individuals are singled out and glorified that should rather be seen embedded in the context of the historic development. Much luck and coincidence is needed to be successful and be selected. Prizes can hardly do justice to those brave men and women who devote, in an unselfish way, all their efforts and energy towards a goal that is finally reached by others.

I am one of the very fortunate scientists who have achieved what many claim to be the ultimate form of recognition or even the ultimate form of happiness in this exuberant, splendid, almost unearthly setting. However, I think more important is the responsibility that is being loaded on the shoulders of the laureates who are supposed to suddenly behave like unfailing sages although they might have been just work addicts in the past. The disproportionate importance that is attributed to the Nobel Prize is reflected also in disproportionate expectations from the public.

Recently, I got a set of letters, written by school children from Bedford, Massachusetts, one of them begging me to work hard towards an artificial ozone layer to protect life on earth. I hope that I can live up to a few of these very high expectations and I ask you already now for indulgence in your future judgments. With this hope, I would like to close and to thank you for your very kind attention."

After the speeches and toast to the King, the strict protocol demanded that the King and Queen stand from the table and take the first dance in the Prince's Gallery. Even Magdalena and I took to the floor for a few steps, the first time we had danced for a long time. All the formal etiquette was a little too much for our daughter Anna, who had never enjoyed dancing. Even at school events, she preferred to stay hiding under the table rather than take to the dance floor. On this evening, she also turned down the opportunity to come with us into the ballroom. Instead she opted to take a little fresh air in the cool Stockholm night, where she spent a good while talking to our chauffeur for the evening, who– it turned out – was actually a psychoanalyst in his day job. She later told us that she enjoyed this part of the evening far more than all the pomp and ceremony.

It was a grandiose evening, which many laureates before and after me have also had the chance to experience. Vladimir Prelog, the originally Yugoslavian ETH chemist, who had received the Nobel Prize 16 years previously, wrote me a humorous letter of congratulations, in which he revealed that even he had been extremely impressed by the prize ceremony – and this from a man who eschewed convention at every turn! It is said that he was the first professor at ETH to wear white turtleneck sweaters instead of a shirt and tie, yet even he wore full evening dress to the Nobel Banquet.

The Nobel Prize winners of 1991. From left to right: Ronald Coase (Economics), Richard Ernst (Chemistry), Erwin Neher (Medicine), Pierre-Gilles de Gennes (Physics), Bert Sakmann (Medicine), Nadine Gordimer (seated, Literature).

After the seven-hour marathon celebration, Magdalena and I were finally able to retreat to our hotel room. I was finally able to get out of my dress suit, in which I had felt rather uneasy. It was the first and most definitely the last time you will ever see me wearing such a thing. Even though I would never turn up in my lab wearing a turtleneck sweater, this was all a bit much for me. But the evening wasn't yet completely over: The journalists who had accompanied us surprised us with a bottle of champagne, a birthday cake and flowers for Magdalena in our room. So we closed out this marvelous day, 10 December 1991, by celebrating my wife's birthday. At the time, I exclaimed that "the journalists have saved my marriage", because unusually, I had completely forgotten what day it was!

Being awarded the Nobel Prize was without a doubt the highlight of my life. Yet this still did not mean that I had laid all my self-doubt to rest. Even today I still feel that I didn't entirely deserve the honor. I've even had dreams where I have received a phone call notifying me that they are taking the prize back off me. Joking aside, these are painful moments that I still can't fully rid myself of. But ultimately, these are outweighed by happiness, which is also shared by many others, thankfully mainly by my family, my friends, and the many loyal, selfless scientists I have been lucky enough to work alongside over my career.

The Nobel Prize medal is made of gold. Richard Ernst had thought once he had lost his medal. But he finally found it again and bequeathed it to the archives of ETH Zurich.

Thangkas: The Other Dimension

First contact

The story of my passion for thangkas began much earlier, back in spring 1968, after five successful years in California. Once Magdalena and I had decided that we would return to Switzerland for good, we sent our daughters Anna and Katharina back to my mother on a flight from San Francisco accompanied by a close friend of ours. We, however, were going to take a detour via Asia. We spent a little more than a month traveling through the Far East, visiting Japan, Hong Kong, Cambodia, Thailand and India, before arriving in Nepal, where we wanted to see (but not climb!) the giant Himalayan peaks around Mount Everest. But when we landed in Kathmandu, it was raining, the skies were gray, and there was no sign of the mighty mountains we knew were hiding in the clouds. So instead we decided to visit some Tibetan monasteries, one of which was Swayambhunath, where we were lucky enough to witness a wedding ceremony.

The old town in Kathmandu was an unbelievable visual feast, with its tiny shops and narrow alleyways, all the beautiful houses with their carved facades. "The almost medieval views were still unspoilt by cars and western tourists," we noted in our travel journal. We soaked up all the picturesque scenery, women with countless nose-rings and earrings. Together with our guide, Narendra Shakya, we ambled through the market and, much as we had done before in Hong Kong and Tokyo, we browsed the art and antique shops searching for treasures. It was in a small shop on New Road that we discovered an abundance of Tibetan paintings, so-called "thangkas", which depicted religious and monastic scenes. As it later transpired, this was an eye-opening experience.

We purchased a bronze figurine and two thangkas. At the time, we had no intention of collecting these valuable paintings; we simply wanted a few souvenirs to take back to Switzerland with us. The items we bought were also cheap enough that a scientist starting out in his career could afford them. The first thing that particularly struck me about the thangkas was their aesthetic appeal: The boldly colored imagery, the masterful and detailed design, the expressions of the figures in the paintings, often embedded in stylized, heavenly surroundings among nature or in a monastery. I found it all simply enchanting. At the start, however, I had no appreciation of the spiritual power of the paintings and my scientific background was no great help in gaining a deeper understanding of the artworks.

"Thangka" is the Tibetan word for "scroll" and the images are painted on cotton cloths, sometimes also on silk. They are then stretched between a bar and an artfully carved wooden pole, on which the paintings are rolled up for storage and transportation. In monasteries, the monks get them out, unroll them and then view them while falling into a meditative state. Thangkas are also often held and displayed during processions. They are deeply embedded in Buddhist culture, which first took root in Tibet in the 6th or 7th century AD, having come from India. There are also similar traditions in Indian Hinduism as well as the Bon culture from pre-Buddhist Tibet. A common feature shared by all thangkas is that they depict figures from the Buddhist world of deities, saints and lamas (spiritual leaders similar to gurus in Indian Buddhism) and often show them in natural surrounds.

One of the two paintings we came across in the Kathmandu marketplace on our first trip to Nepal was of particularly high quality and immediately captivated our attention. It originated from the 18th century and depicted four Buddhist saints, known as Arhats, surrounded by lotus flowers and peacefully flowing streams, painted in the Chinese style. The silk surround of the painting was still in good condition, as was the wooden bar. We purchased our first two thangkas for what we considered a very reasonable price of a few hundred dollars and we were enchanted by our "haul". I remained in contact with our guide from that trip for many years, and he would always send me pictures of new thangkas, but none of them were of the same quality as the first.

Back in Switzerland, however, I had more pressing concerns. I was coming across so many obstacles at ETH that my problems and failures were beginning to pile up. My scientific career was getting bogged down, and in spring 1970 I suffered a nervous breakdown in Winterthur. I thought that I was reaching the end of my productive years in research and feared that I would have to take a back seat for the rest of my life. My doctor prescribed me a period of rest and recuperation, which I spent together with Magdalena in Ticino, southern Switzerland. I put my work aside, while we took relaxing walks and recovered. So I found myself, during the greatest crisis I had faced in my life to date, with time to ponder my own thoughts.

One day – by which point I had recovered somewhat – we were browsing souvenir stores and shops in Lugano's pretty old town. And then fate struck. It was in a touristy cuckoo clock shop under the arcades on Via Nassa where I discovered two thangkas and a gilded bronze figurine of Avalokiteśvara, a deity embodying compassion. Legend has it that this deity once saw the suffering of humankind and his head split into eleven pieces out of sorrow. These then formed into eleven heads, with which he was to hear the cries of the suffering. Nine of the heads have peaceful expressions, but the tenth head has the fearsome coloring of Mahakala, the protector of the Buddhist teachings, the so-called Dharma. At the top sits Buddha, the all-transcending sage, on his throne. This statue and the two thangkas were carelessly left lying between a variety of cuckoo clocks and cheap antiques. It was likely that they had only recently arrived in Europe, probably brought across by Tibetan refugees in the major exodus of the 1960s.

One of the thangkas, which cost us several hundred francs at the time, was amazing; I had hardly come across anything of the like. It depicted Yamantaka, also known as the "conqueror of death", a dark blue, fearsome deity with nine heads. Eight of these heads look at you with expressions of excitement and anger, while the top head has a more peaceful expression (see p. 177). It represents the bodhisattva Manjushri, the "enlightened being" of wisdom. Manjushri is often depicted with the book of wisdom in his left hand and a sword in the right, which he uses to cleave the clouds of ignorance. Later, it was as if the scales fell from my eyes when I realized just how much this image also represented the essence of science. The fearsome Yamantaka ("he who destroys Yama") can be seen as a metaphor for a scientist, who requires the power and tenacity of a wild deity as well as the benevolent and everlasting wisdom of Manjushri in order to succeed. In a way, scientists are also constantly striving for some sort of immortality. They are searching for the ultimate discovery, they want to find the immortal truth, they are hoping that their discovery will ultimately outlive them, if possible for centuries to come. Producing an everlasting achievement was also one of the fundamental motivations for me as a scientist. So it appeared to me that this mystical figure, in an enigmatic way, was a representation of my own ambition. The scrolls featuring Yamantaka are some of my favorite thangkas. The painting we bought all those years ago in Lugano still hangs in our living room in Winterthur.

Yamantaka also symbolizes the core of eastern religions, and the Buddhist philosophy in particular. He symbolizes a connected world, in which good and evil are merely two aspects of a common deep truth, and in which the two are not diametrically opposed, as is the case in our western world. And Yamantaka is no exception: Divine beings and deified historical figures are frequently depicted in Tibetan paintings, but also in bronze Buddha figurines, as having both fearsome and peaceful aspects at the same time. Here, the artists try to bridge the gap between the conflicting aspects of life, with the aim of finding a peaceful middle ground.

The other thangka we bought in that Lugano shop cast a spell over me in a different way. It showed Buddhist monastic life in such fascinating and vivid detail – I could barely take my eyes off it, despite the wear-and-tear it had suffered over the years (see p. 178). The longer I stared at the painting, the more details I discovered: Monks paying homage to a lama, a studio in which figures were keenly painting new thangkas using traditional methods, alongside a monk doing nothing, simply sitting there and staring out into the landscape – it was like one of those films that you can rewatch on many occasions without getting bored in the slightest.

I was quickly impressed by this fresh approach adopted by Tibetan artists, their sense of spontaneity and the metaphorical power of their art. While we had bought items of medieval art before, such as Russian icons mostly representing saints, we preferred the thangkas, as the scenes or the Buddhas and masters portrayed in the works told a real story, throwing light on their history. In addition, it is impossible to separate Tibetan art from Buddhist philosophy and spirituality; all of these have been passed on from masters to students over the centuries.

A culture spanning the millennia

Buddhism reached Tibet from India in the 7th century, gradually replacing the previously widespread Bon religion. Despite its geographically remote location, key trade routes ran close to Tibet, transporting cultural and economic goods along them. The Himalayan passes were much-frequented, albeit extremely inhospitable transit routes from Nepal and India into the northern plains, to the holy Mount Kailash as well as the salt lakes of Changthang. At the time, Tibet was far bigger than today, stretching from Afghanistan and Turkestan in the west to Xi'an in the east, the Bay of Bengal in the south and South Mongolia in the north. The country was subject to absolute rule by King Songtsen Gampo (604–650), who made Buddhism the state religion. Two hundred years later, followers of the pre-Buddhist Bon religion revolted and almost completely wiped out Buddhism in Tibet. This ancient religion, whose believers also used to traditionally produce thangka paintings, found a new lease of life until the 11th and 12th centuries. During this period, a second wave of Buddhist teachers arrived in Tibet, partly driven by the Islamic conquests in the Indian sub-continent. The most well-known scribe from this period was Atiśa (980–1054), who translated Buddhist texts from Sanskrit into Tibetan. A great many monasteries were then established in central Tibet, which would go on to shape cultural life there.

Nowadays, Tibetan culture appears to be very homogeneous; it is Buddhist through and through. Even everyday tasks are carried out in accordance with the requirements set out in Buddhist teachings. In essence, Buddhism is an atheistic religion. Buddhists do not recognize the existence of a creator deity that exists outside of his own creation. Everything in this world is interconnected. Even when a change occurs, this is embedded in a recurring cycle that the universe goes through. Everything we do leaves its traces, which means we have an immense responsibility.

Despite its atheistic leanings, Tibetan Buddhism actually features a whole pantheon of deities (Buddhas) – from divine protectors (Dharmapalas), yogis (Mahasiddhas), saints (Arhats), through to deified figures (Bodhisattvas, or people on the path to Buddhahood) and accomplished masters, all of whom are on the Buddhist path to enlightenment. There are several hundred of these figures in the Buddhist faith, with many variations being found depending on the region or era. They are characterized by their number of heads, arms, legs, their posture or position of their hands, as well as their color. It therefore comes as no surprise that this vivid universe of figures has provided fertile ground for the imaginations of painters and sculptors over the centuries.

The monasteries founded during the first and second Buddhist waves produced countless thangkas as a means of edifying visitors and to help the monks in their meditation. They created the paintings in monasteries' own art schools or commissioned them from outside workshops, which were often run by well-known masters. As is the case for Christian orders, the monasteries were split into different schools. There is the Nyingma school, which was founded in the 7th century, and the Kagyu, Sakya and the reformist Gelug school, which followed later. These schools were founded by extraordinary saints who had traveled the path to enlightenment and had translated the Buddha's teachings. Many sub-schools also emerged along the Buddhist lineage, which always passes from the masters to the scholars. These also developed their own traditional styles in the art that they produced, which is nowadays extremely helpful when it comes to dating paintings, and also of great historical interest as it reveals the lineage of the school of faith and its line of descent from Buddha.

Many thangkas were commissioned by monks in the monasteries, where they were carefully stored and maintained for centuries, meaning there are still fine examples from the 11th and 12th centuries in surprisingly good condition. However, the brutal Chinese invasion in 1950/51 and Mao's devastating Cultural Revolution lasting between 1966 and 1976 caused a huge amount of suffering and destruction in Tibet. Monasteries were torn down, while thangkas were stolen and destroyed. The Chinese also expelled many spiritual leaders and monks, who fled to Nepal and India. Alongside their belongings, many also took a thangka with them, which they would sell to antique dealers in the streets of Kathmandu or New Delhi so they could afford basic food. This resulted in the art market being flooded with these wonderful scrolls, although at first, they largely went unnoticed. This is certainly no longer the case.

Prices have since shot up to astronomic levels, with some thangkas today selling for hundreds of thousands of dollars. But still almost no paintings from the monasteries of Tibet make it to the West as they are now classed as strictly protected cultural property.

A joint passion for collecting

Since acquiring our first thangkas in Kathmandu in 1968, my wife and I built up a significant collection; our house used to sometimes be bursting at the seams with Asian artworks. Whenever on our travels, we would spend time in antique shops and galleries on the lookout for these fabled scrolls. When I won the 1991 Nobel Prize for Chemistry, with its prize money totaling more than one million dollars, my collection got a real boost. I invested a significant amount of this money, which I had earned through all my hard work, in precious thangkas. I could finally buy what my heart truly desired. It was not long before our house was covered in Asian art – you could find it in our living room, in corridors, above our bed, in our offices. Anything we could not put on display, we stored in the best possible conditions in our cellar, hallway, or garage. Even our car had to make way for thangkas.

Right from the start, our collection was a joint project. When we recently sold the majority of our collection at auction at Sotheby's, the sale was held under both our names: "The Richard R. & Magdalena Ernst Collection of Himalayan Art". Why "Himalayan" and not only "Tibetan" art? Over the years, we had expanded our collection to include Mongolian thangkas, Asian book printing and – to a lesser extent – statues, as the prices being demanded for Tibetan thangkas had reached horrendous levels.

However, Tibetan thangkas were and still are the source and indeed the heart of our passion for collecting. The first thangka we bought in Kathmandu depicted four Arhats, in surroundings so harmonious and so pure, that as soon as I got home I knew straightaway that this first painting would soon expand into a collection of some sort. The skies on the painting were also lit with beautiful white clouds and shortly after returning from our travels, we came across the book "The Way of the White Clouds" by Lama Anagarika Govinda – this had to be a sign! Lama Govinda was born in Germany and was an archaeologist and philosopher, who later acquired British-Indian citizenship. In the 1930s, he embarked upon his journey as a Buddhist and spent many years traveling through India and Tibet, where he learned about the various schools of Tibetan Buddhism, which he described in his 1966 work. During the Second World War and around the time of Indian independence, he was interned by the British for five years as a result of his association with Mahatma Gandhi's movement. His books and writings helped shape the image of Tibetan Buddhism in the West. It was these impressive texts that strengthened my and Magdalena's resolve to build up a collection of Tibetan thangkas. Lama Govinda's vivid explanations allowed us to view everything we had experienced in Kathmandu and everything we had seen in the paintings in a spiritual light, something that we would not have appreciated so clearly otherwise. This gave our collection a deeper meaning.

I was very much interested in this spiritual aspect, while Magdalena – who had always had a flair for the visual arts – judged paintings on their artistic merits. I found thangkas to be the ideal gateway into this exotic world, given their symbolic language allowed even the man on the street to have an intuitive understanding of them. Each individual painting contained its own universe, opening up new insights into the Buddhist world. This allowed me to discover a worthy alternative to the materialistic philosophy so predominant in the West. Even for a rational scientist, Buddhism is easy to understand given its simple philosophical and ethical rules, which in no way contradict the basic scientific principle I hold so dear. It quickly becomes apparent that the great many deities in the religion act as metaphors for various philosophical principles. As such, religious art gave me access to an unspoken spirituality and religious symbolism, which fed my desire for a holistic view of the world.

In selecting paintings for our collection, we miraculously managed to find common ground. We agreed on one unspoken rule: If we found a painting we both liked, we would go for it. If one of us had reservations, we would let it go. Most of the time at least! There were times when I snapped up a painting despite my wife's objections, and I then usually realized after the event that I should have actually stuck more closely to our "rule". Once in the summer of 1969, for example, I was in a cheap antique shop in the Zurich old town, when I came across a thangka I simply had to have. The dealer wanted around 600 francs for it, which I thought reasonable. My wife advised me against it, saying how you could see from a mile away that it was a poor example of a thangka – dark, crudely painted, in poor condition all round. But I insisted, I just wanted to build up our collection of thangkas! Ultimately she bought me the painting for my birthday. So without technically breaking the "rule", I got my thangka after all. But while the values of the best thangkas have skyrocketed over the last forty years, this particular example is barely worth a thing nowadays – I think it's still lying around the house somewhere, having never even been hung.

Nevertheless, our collection was never about the money. We bought the paintings we liked, only considering a purchase if it "touches our hearts", as Magdalena likes to say. There were no strict criteria we applied when buying a thangka; we let our experience and intuition lead the way. In parallel to building up an ever more impressive collection, we gradually immersed ourselves in the historical and philosophical aspects of this art. We read books, researched the roots of Buddhism in Tibet, learned about the symbolism and imagery in the paintings, and I also became extremely interested in deciphering Tibetan inscriptions. We soon also started crossing paths with a number of famed Tibetologists and collectors. One of these was the well-known New York collector Jack Zimmerman. Jack and his wife Muriel had built up one of the most significant collections of Indian and Southeast Asian art, with their works featuring in numerous exhibitions and publications. Sadly, Jack recently passed away, but our first meeting was telling. It turned out that one of the thangkas we had bought in Lugano was extremely special. It was so rare because unlike most other thangkas, its main focus was not a Buddhist deity, but instead it depicted life in a Tibetan monastery in great detail. Soon after buying it, we lent it to the Kunsthaus in Zurich and the renowned Musée Guimet in Paris for their exhibitions. When Jack saw the painting there, he liked it so much that he decided he just had to meet its owners. So one day, he made his way to Winterthur, where we gave him a warm welcome. When walking round our house, his attention was drawn to another scroll hanging in our bedroom above our bed. It was painted in the Hindu tradition and not the Buddhist, a fact that Jack noticed immediately. His eyes lit up: "That's my thangka!" he exclaimed with a smile. It transpired that he had recently sold this very same painting to a dealer in Zurich, from whom we had then bought it. It was as if the god of thangka collectors had unconsciously bound us together with an invisible thread. It also marked our entry into the extensive community of Asian art collectors. From then on, we maintained a friendly relationship with Jack and many other interesting characters who had – for one reason or another – felt drawn to Tibetan art.

Journey to Tibet

In summer 1993, we were finally able to travel to Tibet – it was to be the first and last time that were able to experience this secretive and fabled Himalayan region in person. At the time, the country had been largely sealed off by China, with only handfuls of tourists able to visit. It was not until later that China opened up Tibet's borders for western tourists, and even then only in a strictly controlled manner. Our trip had been made possible thanks to my work as a consultant at Bruker, which at the time already had a branch in Beijing. Arriving in Beijing, we were accompanied by an extremely friendly and attentive guide, who tried to engage us in conversation about all manner of subjects. I got the impression she was impressed by me, perhaps having even heard about my scientific achievements, something that may well have made a real impression in a country so hungry for knowledge as China was. Our guide – unfortunately I can no longer remember her name – even organized a big party for my 60th birthday, which I celebrated at the start of our trip.

From Beijing, we then traveled to Chengdu, before moving on to Lhasa. The attention paid to us by the Chinese officials assigned to us during our stay in Tibet was like nothing else. The authorities had given us a guard; I called him "the gorilla". His job was to accompany us every step of the way, trying to prevent us having any contact with the local population, or at least recording any contact we might have. I was unable to conceal my real annoyance at this, something the pitiful official clearly picked up on. After a short while, he climbed exasperated out of our car and let us continue on our own. This came as a real relief and a weight lifted off my shoulders. Now we could really get going. We drove into the mountains, first into the town of Gyantse, which is located on the Friendship Highway at around 4,000 m above sea level. With a population of 40,000, Gyantse is the fourth largest town in Tibet, but it is mainly known for its Pelcho Chode monastery originating from the 14th century.

Most monasteries in the Tibetan heartland were destroyed and looted during the political turbulence with China, especially during the Cultural Revolution at the end of the 1960s. Even though many have since been rebuilt, they lack the splendor and aura of centuries gone by. They are now merely functional buildings for monks to live in and pray. But this is not the case in Gyantse: "The Kumbum was spared from the destruction wrought during the Cultural Revolution by Zhou Enlai, as was the Pelkhor Chode," noted Magdalena in our travel journal. Zhou Enlai was a pragmatic statesman during Mao's revolutionary reign and spent many years as foreign minister and premier of the country. He prevented the Red Army from destroying this monastery, despite many other sites falling victim to destruction. "It is hard to believe that so much valuable Tibetan art still remains," wrote Magdalena in her journal, describing our sense of surprise. At the center of the site lies the Kumbum built in 1497, a giant reliquary in the form of a multi-level pagoda. With nine floors, 108 chapels and tens of thousands of wonderful wall paintings, the Kumbum is one of the most architecturally important sights in Tibetan Buddhism. Like a three-dimensional mandala, its architecture is based around a circle within a square. We marveled at the many thangkas hanging in the holy chapels and hallways.

We went on to visit the famous Ngor monastery. Also founded in the 15th century, it was the cultural center of the Sakya tradition, one of the four main schools of Tibetan Buddhism. During its early years, it was renowned for its monks' and priests' love of art. The monastery itself looked miserable – it was one of those sites which had been rebuilt following the turbulence of the Chinese invasion and the subsequent Cultural Revolution, but in a crude, functional manner. Far more affectionate and open was the contact we enjoyed with the locals. The monks invited us into the monastery and talked to us about their sorrows and joys. We often spoke about the Dalai Lama and on occasions, gave them pictures of their worldly, spiritual leader, which we had brought with us from home. Amazingly, these pictures were extremely sought after, because under Chinese communist rule they were strictly forbidden and in Tibet itself, impossible to find. When it was time to leave the monastery – I read in my journal – we wished the monks a long and happy life.

In Tibet itself, collectors will be disappointed if they are looking for valuable thangkas. Many of them were destroyed during the Cultural Revolution, and in the Tibetan heartland, there are no longer any workshops producing paintings using traditional methods. Any tourist markets are full of cheap, second-rate thangkas from Nepal or India. Nevertheless, there are still many wonderful thangkas originating from the Ngor monastery that can be found on the art market. Indeed, I found out some years later that the two thangkas we had bought in that cuckoo clock shop in Lugano back in 1970 actually originated from here.

The Ngor monastery was also a center for Mandala artworks. It produced a particularly valuable thangka in our collection, which I bought at auction from Sotheby's in 1999 for almost 70,000 dollars: The Kalachakra mandala (see p. 180). Mandalas are a real firework of symmetries, but also contain broken symmetries. While traditional thangkas show scenes from the lives of the enlightened, depict masters or also their students, Mandalas feature hidden lines of force, which influence the spiritual universe. In this respect, Mandalas are similar to modern science, for example when particle physicists or cosmologists are looking for the last secrets nature has to offer and the laws by which they are bound.

The Kalachakra mandala aims to correlate space and time, and it features beautifully crafted, minute details. Its real depth only becomes apparent after you have spent many patient hours observing every detail, even using a magnifying glass as necessary. The viewer – most probably a monk on the path to enlightenment – moves gradually from the outside to the center of the painting, which contains a revered deity representing the theme of the mandala itself. The Kalachakra mandala represents the "wheel of time". At its center is the deity Kalachakra, with his 24 arms and four heads, embracing his female consort, the eight-armed Vishvamata. There are no limits to its symbolism: The outer ring of fire represents the eight mystical Indian cemeteries, while the circle is split into four differently colored quarters, symbolizing the four elements: Earth, water, air, and fire. The inner area of the circle contains four concentric squares, with large gates on each side, through which the meditating monk has to symbolically pass, so as to slowly make their way into the center to the revered deity.

"Once a chemist, always a chemist"

Between 1991 and 1996, after I had won my Nobel Prize, I spent around 900,000 dollars on thangkas. This included some exquisite items, which would go on to increase significantly in value. At the start of 1992, for instance, we acquired a particularly valuable thangka depicting four masters from the Kagyu school from famed art dealer Chino Roncoroni. I will come back to this thangka later on to show in detail just how complicated it can be to analyze an artwork of this caliber.

Another thangka bought during this period is my wife's favorite. It comes from the 15th century, is exquisitely painted, and represents the deities Guhyasamaja and Sparshavajri (see p. 181). Guhyasamaja can be seen as the embodiment of hidden union, while Sparshavajri is his female counterpart. The painting shows these two six-armed deities in harmonic union, closely embracing. Both hold the same symbols in their hands and are adorned with valuable gold and jewelry. In a certain sense, the painting also reflects the philosophy that in the West is mainly known as "yin and yang".

We acquired many of our paintings at auction, while others we discovered in galleries or even on the internet. Sometimes buying such valuable pieces of art was a nerve-wracking experience. It was often the case that you had to form an opinion about a piece within a very short space of time if you wanted to buy it. This is no easy task, and it often pushed our unwritten rule – that we would only buy paintings that we both like equally – to its limits. Is the painting genuine? How old is it? Which school is it from? Beyond their significance with regard to artistic and historical merit, the answers to these questions can also have a direct influence on the price. The difference between a thangka painted in the 12th century and one produced in the 15th century can quickly run to hundreds of thousands of dollars.

In the late 1990s, I indulged my passion for collecting these fine artworks even further. I would spend every free weekend working on Tibetan scrolls, as confirmed by my diary entries from the time: "Sat., 18.2.1995: Painting and restoring thangka; Sun., 24.2.1995: Looking for identification of Lama thangka; Sat., 25.3.1995: Painting and restoring thangkas; painting, stitching frames; Sun., 4.6.1995: Restoring thangka." I had built myself a small lab at home for restoring thangkas. At first, this was not so much about maintaining the quality of our collection, but was more a case that thangkas coming onto the market were often in a poor condition. This was also related to the way these artworks were traditionally stored and used. When unrolling and rolling up a scroll, it is common for them to split or lose coloration. With the scrolls that hung for centuries on the walls of monasteries, you could see traces of the butterlamps in use everywhere in Tibet. It was sometimes even the case that the entire canvas would be sprayed with butter.

Richard Ernst learns the art of priming a thangka in a painter's workshop in Kathmandu, Nepal. Photo taken in 1997.

Collectors are therefore faced with the never-ending challenge of not only having to clean and restore their artworks, but also to maintain them on an ongoing basis. This required equipment to examine the chemical properties of the paintings' surfaces using a combination of microscopy and spectroscopy, precise tools and tweezers to clean paintings that had attracted dirt over time, and paint brushes and color pigments to repair minor or even major damage to the layers of paint or the Chinese silk brocade frames.

Conserving works of art is one of the most challenging tasks you can face in the field of materials science. For me, it above all represented a return to my roots, to my original vocation as a chemist, and I enjoyed the challenge very much. I had to dust off all my knowledge of canvases, read up on questions relating to the aging properties of rare pigments, and learn how to properly treat the layers of paint. Applying individual pigments requires a special process for them to adhere to the layers of paint below. Before doing anything, you need to carry out a thorough examination to find out as much as possible about the materials and the conditions they are in. I had to learn how to identify the pigments so that I could then find the original colors in order to repair any damaged areas. At first, my chemistry and analytical knowledge from the first semester of my undergraduate degree were sufficient in this regard. For the first time in thirty years I was able to single-handedly perform chemical reactions. In 2012, I authored an essay titled "Once a chemist, always a chemist!" (*Ein Chemiker bleibt Chemiker*) for the renowned *Angewandte Chemie* journal, which discussed my work in the field of art history.

Despite this new passion, however, I still occasionally had serious doubts about the point of my enthusiasm for collecting art. "I'm actually very unhappy with myself," I wrote on 1 January 1996 in my diary, as I tried to take stock of my life. I was under a lot of pressure at ETH, having to deliver a colossal amount of lectures. Winning the Nobel Prize – something that should have actually made me happy – weighed heavily on my shoulders. Rarely a week went by when I wasn't traveling somewhere in the world to give a lecture or attend a conference. And in addition to all this, I still had a research group to run. But I was troubled by doubts in all areas of my life, also with regard to my passion for Tibetan art, which was supposed to provide me with some balance in my life. "What have I actually achieved in Tibetan art?" I asked myself back then.

There was no easy answer. I tried to come up with a list of pros and cons: "My interest in a non-European culture means a huge amount to me. But does that mean I should keep collecting and investing money in it?" I would have happily spent even more time learning about the secrets of the thangkas as well as Himalayan history and culture, but my work at ETH, my involvement in various research policy committees, and above all, my ever more consuming lecturing obligations around the world all prevented me from doing so. At the time, building and maintaining my collection was therefore "the only opportunity to indulge my active interest in Tibet; I just don't have time for all the other activities," I concluded. "So, I'll keep collecting, albeit more prudently and sensibly!"

When I retired in 1998, I finally had time to delve deeper into the thangkas. There was no stopping me. I expanded my restoration lab into a proper laboratory for conserving art. I got hold of a stereomicroscope, together with all the lenses, polarizers, and photographic equipment I needed. This sort of microscope is used in performing experiments, but is also vital when it comes to restoring and touching up layers of paint. I mounted it on a mobile rig, which allowed me to place paintings up to two square meters in area underneath it to be examined.

It turned out that the best method for analyzing color pigments was Raman spectroscopy. It is based on the ideas of Indian physicist Chandrasekhara Venkata Raman – better known as C. V. Raman – a brilliant scientist who broke into the phalanx of the European intellectual giants at an early age, and was awarded the Nobel Prize for Physics in 1930 aged just 42. This method involves using a weak laser beam to scan the entire painting. Raman discovered that this produces scatter radiation, which varies depending on the material. This means that each pigment creates a very specific spectrum and by using the relevant detectors, it is possible to collect and analyze the scattered light. This turns the spectroscope into a microscope, which can be used to identify the various pigments on a painting. A modern Raman microscope has a resolution of a few micrometers. The main advantage, however, is that unlike with conventional microscopy, there is no need to scrape the pigment off the painting and treat it specially. After all, even if you have to take just the tiniest sample, you are still interfering with a piece of art.

Some years later, in 2007 or 2008, I acquired a modern Raman microscope, which was perfect for analyzing samples in a small lab like my own. I spent many hours at home, building a mobile platform on which to mount the device. Now all I had to do was place the damaged part of the painting underneath the microscope and I was able to analyze the color pigments. This allowed me to examine even huge paintings in high resolution. In the field of art history, the method is frequently used to examine frescos, for example, but to use this method with Asian paintings, in particular Tibetan thangkas, was almost unheard of – I was a pioneer in this regard. This also led to me being able to successfully redate a number of thangkas, work which I later published.

An analysis of colored pigments tells the story of a painting in a completely new way. A Raman microscope shows the exact composition of each color applied to a canvas. With a Buddha, for example, which is traditionally always painted blue, you can see whether it is actually azurite, indigo or Prussian blue – all different tones of blue with their own different stories. Azurite is a copper-based mineral, which has been extracted from the copper mines in Nyêmo, Tibet for hundreds of years. Indigo comes from India, which in historical times was as much known for producing indigo as it was spices. Prussian blue was not invented until the 18th century in Europe, before it later found its way to Tibet.

Some time ago, I also analyzed the thangka of the four Kagyu masters, which I had paid a lot of money for and was particularly close to my heart (see p. 182). In addition to the four masters from the Kagyu school, the painting also featured smaller representations of six of the most important figures from early Buddhism, including Buddha himself in the top center, or Atisha, the historic founder of the Kagyu school. This painting alone is the subject of four research projects, in which renowned Tibetologists and art historians have expended a great deal of effort analyzing the work. The identity of the main figures, however, was still the subject of some dispute – each researcher came to a different conclusion.

Based on the stylistic elements of the work and their knowledge of art history, the experts presumed that the work originated from the 12th or 13th century, most likely from a monk in one of the most significant monasteries of the Kagyu school. As figures on thangkas are depicted in a strictly hierarchical order, the figures on the right should be the masters and the figures on the left, their students. Their personal characteristics are extremely subtle, for example the shape of their heads, the color of their hair or style of beard, all of which provide clues to their identity. This means that even without an inscription, a thangka can be read in almost the same way as a written text.

But with this thangka, it was difficult to identify the main figures, despite the experts being able to consult various well-known portraits from the relevant period. The painting could just as well have been a cheap copy produced relatively recently. The only way to resolve this was to now take a scientific approach, with one option being carbon-14 dating. This involves measuring the concentration of the radioactive isotope carbon-14, which has a half-life of around 5,730 years, thus making it possible to determine the approximate age of a painting. To make use of this method, I used my connections with ETH. I sent Georges Bonani at the ETH Institute for Particle Physics a sample of the cotton canvas from the painting. The results of his analysis showed that the cotton used was harvested in around 1229, albeit with a margin of error of around 70 years. Nevertheless, this showed that the art historians' assumptions about the age of the painting were not vastly different from those based on scientific analysis.

Lastly, I analyzed the pigments using Raman spectroscopy. The red used to paints the monks' robes was made of cinnabar. The small Buddha figure was painted using indigo, as were other parts of the painting such as the stylized blue stones below. The green of the upholstery was a mixture of orpiment yellow and cinnabar red. By contrast, malachite green was not used at all. Malachite green is a copper-based mineral pigment, which like azurite blue, has been extracted from copper mines in Nyêmo, Tibet for hundreds of years. The heavy use of indigo and absence of malachite suggest that the thangka was painted by a Nepalese artist on behalf of a Tibetan monastery, as this range of colors was commonplace in Nepal. At the time, it was often the case that Nepalese art schools would offer their services, as the Nepalese people had a more deeply embedded painting culture than Tibetans. One thing that still puzzled us, however, was the white pigment used in the protagonists' eyes – nobody had any idea what it was. This was not entirely unexpected and related to the mystery surrounding eyes in these paintings. While in the large art schools, students would often paint the thangkas almost to completion, it was always the masters that would finish the work by painting the eyes. It was these final brushstrokes which brought the painting to life, so to speak. But by performing an in-depth Raman analysis, I was able to establish that the white pigment was in fact anatase, a pigment that was not available until 1916. I could only speculate how this relatively modern pigment made its way on to this centuries-old painting. It is likely that the eyes were retouched at some point in the 20th century using this synthetic pigment.

There can be no doubt that scientific methods have really given a new dimension to research into Tibetan art, and I am proud to have played my part in this. I have sometimes come in for criticism from classical Tibetologists, usually when it turned out that a picture was a lot more recent than originally thought. If a Buddha is painted with Prussian blue, for example, it can be assumed that the painting is more recent. For collectors, finding out something like this can mean their painting is worth up to several hundred thousand francs less than they thought. Of course, pigment analysis on its own is not enough to draw watertight conclusions, and the expertise of art historians and Tibetologists remains essential. But when combined with traditional approaches, scientific methods can help us take another step towards discovering the truth.

One of the most significant collections of Tibetan art

I used to often ask myself why I was able to spend so much energy on my art collection alongside my scientific career. At first, I considered my entry into this fascinating world to be that of a secular chemist into a paradisaical world. I was neither a Buddhist, nor was I religious, but my experience of Tibet and its culture gradually expanded my world view as well as my understanding of spiritual concepts. I found that unlike other monotheistic religions, Buddhism did not fundamentally conflict with modern science. A Buddhist is free to draw on any source they wish provided it furthers their wisdom, and this can include rational science. I soon identified a number of similarities. While in chemistry we use formulas to represent structures and chemical reactions, in the spiritual world metaphors, symbols and rituals are used as a universal expression of philosophical concepts outside of any one specific language. My scientific techniques, which extended beyond the boundaries of previous art history approaches, simply increased my passion for the subject no end. It gave me a positive feeling, knowing that I was able to combine two such primeval activities – science and art – and in doing so, improve my understanding of both.

As I have said before, scientists need more than one leg to stand on in order to move forwards. This is something I have always told my students. Getting involved in something outside the specialist area where you wish to make your mark not only expands your horizons, but also enriches your scientific work. For some people this is sport, while others like to read or play a musical instrument. And for me, collecting exotic artworks really became a passion. I am completely aware that this is in part down to my family heritage. In "Winterthur society", collecting art is considered a *raison d'être*, it is an inextricable part of a person's social status, as it were. Anybody who is anybody has a collection – and the more valuable it is, the greater the esteem attached. Rich families, such as the Reinharts or Volkarts, and other patrons of the arts have built up magnificent collections that are still revered today. I did not of course set out to replicate these, but in Tibetan art, I found a fascinating new niche, which – at least in the early days – was also affordable.

Collecting things, much like curiosity, is a scientist's primary motivation and is a fundamental part of human nature. Behind both of these activities is the basic need to leave an indelible mark on the world. We want eternal fame, whenever possible as a result of our own creations. But not everyone can be a creative artist, and most people have to draw upon the creativity of other extraordinary people to leave their own respectable legacy. I expect this is how many art collections came about; this is definitely the case for mine. It was by discovering the centuries-old culture of thangkas that I was able to go some way to cementing my own immortality a little.

It means a lot to me that I can give something back to Tibetan art and culture. I would love for this rich culture, so steeped in history, to be allowed to bloom once again, for it to have a future, as thanks for how much it has given me. I saw an opportunity to make a contribution in this respect, primarily through my contacts at the Tibet Institute Rikon in canton Zurich. It is the only Tibetan Buddhist center in Switzerland to be under the patronage of the 14th Dalai Lama and was founded in the 1960s following the influx of Tibetan refugees. I have had close links with the community for many years now and am proud to support the center. I sat on the foundation board between 2004 and 2018, which was a special honor for me, but also an obligation to be taken seriously as it allowed me to advise on the future of the institute. On occasions I was able to examine some of the valuable thangkas which had been donated to the institute as a means to finance their diverse range of activities. I particularly enjoyed having the opportunity to compare and discuss western and eastern ways of thinking, which was possible as part of the "Science meets Dharma" program, for example. In Tibetan Buddhism, the term "Dharma" refers to the teachings of Buddha, and the aim of this program is to discuss how modern science and traditional Buddhist culture can have a mutually beneficial effect on each other. The Dalai Lama himself laid the foundations for this in 1998, when he requested the creation of the Tibet Institute Rikon's own university to mark its 30-year anniversary. Since then, teachers and scientists have regularly visited Tibetan exile monasteries in southern India as part of the program, where they teach the monks about science and scientific thought. But as the name of the program suggests, knowledge flows both ways: the western scientists are able to experience a new, spiritual world through the monks, one which will hopefully help them be more in touch with their inner selves. I would like to think that this exchange of thoughts and cultures will also have a positive impact on science, a field usually synonymous with strict rationality.

Richard and Magdalena Ernst in a circle of Tibetan monks participating in the exchange program "Science meets Dharma". Photo taken around 2003 in a monastery in southern India.

By the end, Magdalena and I amassed a collection of almost 1,000 works. The Richard R. & Magdalena Ernst Collection was rightfully seen as one of the most significant collections of Tibetan and East Asian art, in Europe at least. In 2018, we sold some of our most beautiful thangkas at auction at Sotheby's in New York, as we were no longer able to properly conserve them at our home in Winterthur. The wonderful catalog sold quickly, with some thangkas going for high prices, but each sale pained me, so I am happy that at least some of my favorite scrolls, such as the Yamantaka, are yet to find viable buyers.

Four Arhats. The painting depicts the four Buddhist saints (Arhats) Kanakavatsa, Vajriputra, Ajita and Bhadra (clockwise, from top left) in mountainous landscape with rivers flowing through it, surrounded by lotus flowers and white clouds. Arhats were originally companions of Buddha, 16 in number, who overcame the eternal cycle of suffering and rebirth, and finally reached Nirvana. The top right of the painting features the Buddha Amitayus; he embodies one of the three dimensions in which Buddha is represented. At the bottom right is a worshiper offering a large bowl of fruit to the saints. The thangka was painted in the 19th century in Tibet. Richard and Magdalena Ernst originally acquired it in 1968 in an antique shop in Kathmandu.

Yamantaka. The dark blue buffalo-headed deity wearing precious gems and human bone adornement, represents a Yamantaka – the destroyer of death and a favorite of Richard Ernst. There are numerous Yamantakas in Buddhism, but the eight-headed deity depicted here – Vajrabhairava – is the most well-known. The figure is surrounded by a variety of Tibetan lamas and Indian deities with many arms, sitting in geometric registers around the central figure in a clearly ordered hierarchy and feature characteristic emblems and animal symbols. The figure on the top of Yamantaka represents the deity Manjushri, with a content smile on his face; he is also known as the god of transcendental wisdom. The thangka was painted in the 15th century and combines the power of the destroyer of death with the vision of eternal wisdom.

Monastery Life. This rare and important thangka from the 18th century depicts episodes from the life of the abbot Rinchen Migyur Gyaltsen of the important Ngor monastery in central Tibet. In the upper left, the abbot and two hierarchs focus on Buddhist teachings, while five thangkas hang on the walls. A group of monks gather at the feet of the master in a building complex in the top right of the painting; to the left the master receives tributes of fine silks, furs and sutras from monks and laity. Beneath to the right the eminent artist Shuchen Tsultrim Rinchen supervises students while diligently painting a set of thangkas for the abbot. Only one student is not doing so, instead leaning casually on a balustrade, looking stoically into the distance. Multiple gold inscriptions cover the current thangka, indicating the scenes presented.

Takla Membar. This somewhat loose-form, less structured thangka points to the painting having been produced in a different context to the other thangkas. It is actually a painting in the Bon tradition. Bon is the precursor religion to Buddhism; it was gradually replaced by the latter, but its tradition still lives on in certain parts of Tibet. In terms of the painting's style, there are similarities to the Buddhist tradition, but there are also various iconographic differences: Skull crowns, nine crossed swords, screens, as well as butterlamps and birds. This very carefully composed work originates from the 18th century and depicts the wrathful deity Takla Membar, the flaming tiger god, with a skull crown and flaming head of hair. In his right hand he is holding a golden chakra (wheel of time) and in the left, a weapon of crossed swords.

Kalachakra Mandala. In contrast to deities, the enlightened, monks and laity, mandalas are quite rarely depicted on thangkas. This mandala originates from the Ngor monastery in around 1570. The original painting is relatively small (50 cm wide and 50 cm tall), but is extremely elaborate and detailed. Mandalas are feasts of symmetries and broken symmetries and attempt a comprehensive representation of the spiritual universe. They are centered on a venerated deity and serve as meditational aids in focusing one's attention sequentially on deeper and deeper aspects of the deity, progressing mentally from the periphery to the center of the mandala. The deity Kalachakra depicted in the center of this mandala is a male representation of the wheel of time, embracing its female equivalent Vishvamata. Together they symbolize the unity of movement (Kalachakra) and the perpetually static element (Vishvamata).

Guhyasamaja. This particularly valuable thangka was painted in the early 15th century and symbolizes the unity of opposing poles. It depicts the two deities of unity: The dark blue three-headed and six-armed Guhyasamaja in union with his three-headed and six-armed light blue consort Sparshavajri. The pure red space of the shrine serves to highlight the dynamic line of the monumental figure and subtle blue shading gives dimension to the goddess's lithe form draped with scarves and exquisite gold jewelry. A large number of deities, followers, saints and other important figures in Buddhism are arranged in geometrics registers around the central shrine. All the figures are clearly identifiable through the symbols or from context; they follow the hierarchical lineage of the Buddhist tradition.

Four Kagyu Masters. This ancient painting from the early 13th century depicts two masters and their disciples as the main figures. The smaller figures arranged in the rows above them represent some of the most prominent spiritual leaders in Tibetan Buddhism. Buddha himself is sitting in the top row, while the figure in the middle of the lower row of three is Atisha, an Indian saint, who translated Sanskrit writings into Tibetan and founded the famous Kagyu school, one of the four major traditions in Tibetan Buddhism. On his right is Milarepa, the most important Tibetan poet, whose works are still passed down to this day; as always, he is depicted in white robes. Even now, experts have been unable to firmly identify the main figures. The upper left teacher is likely to be Gampopa, one of the most important disciples of Milarepa and an influential master of the Kagyu tradition. Next to him is his disciple Phagmo Drupa.

Early Lama. This ancient and valuable thangka from the 13th century is almost one meter tall and 72 cm wide. It depicts an unknown master from the period in which Buddhism was finally establishing itself in the 10th to 12th centuries in a second wave from India. Lamas such as the master represented embraced and preached the Buddhist path to enlightenment. As there is no inscription to provide an indication of his name or affiliation with a tradition, it has not yet been possible to unequivocally identify him. It is, however, beyond doubt that he was a highly respected teacher; the two Bodhisattvas – those on the path to Buddhahood – worshiping him almost infer upon him a Buddha-like status. Mythical creatures in the niches below him guard and elevate his seat almost to the extent that it appears to be a throne in the heavenly pantheon of Buddhism.

Episode from the Jataka. The Jakata tale depicted on this rare thangka from
the 19th century relates the legend of Buddha's self-sacrifice in his previous
incarnation as an elephant. Buddha encounters a group of people threatened
by starvation and sends them to a lake at the foot of a cliff. In the form of an
elephant, he throws himself off the cliff and lying on his back, is surrounded
by the starving people, who are then able to feast on his remains. The Jataka
tales, which originated around 2,000 years ago in India and detail Buddha's
experiences, usually including a moral or ethical message. The collection
contains 547 episodes in poetic form; some of them, the "Jatakamala", describe
events in Buddha's early life, in which he was often reincarnated as an animal,
as is the case here.

The Eighth Dalai Lama. This thangka depicts the enthronement of the eighth Dalai Lama, Jampel Gyatsho, in 1762. It is one of the finest examples of Lhasa high court style painting. The Potala Palace is depicted to the left, where the ceremony is reproduced in detail. To the right of the main figure is a small representation of the birthplace of the Dalai Lama, connected by a gold line to his spiritual progenitor above. Animated scenes of adoration portray foreigners bearing gifts, monks with offerings and Lhasa nobles in traditional costume. Gods and guardians mingle with the people, symbolizing the dual spiritual and political role of the Dalai Lama. In the clouds above the scene, the Dalai Lama's teachers and predecessors overlook and protect the enthronement.

Legacy

The parable of the three forgotten words

Diary entry, Monday, 1.1.96: "8:00, woke up / thoughts about the new year: I'm actually very unhappy with myself. What have I achieved recently? At work? In public life? In my family? In Tibetan art? Despite my privileged position, the results of my efforts are truly pathetic."

I began to keep a diary at some point in the middle of 1995. If you are fairly disciplined in doing so and write down what you are feeling every day, it means you don't have to remember irrelevant details. I was always blessed with the gift of a poor memory; a lot of my experiences and thoughts would soon disappear into a fog that quickly blanketed the past. I also found, however, that keeping a diary was somewhat akin to going to confession in church: Writing down your feelings and experiences takes a weight off your shoulders, which you no longer need carry around; it alleviates any feelings of guilt you may have.

Even before I began to commit my thoughts to paper, I knew that I would not be happy with the results. My entry from New Year's Day 1996 shows that I was once again dwelling on my imperfections and was rarely willing to accept my own limitations. It seems this was my eternal destiny. Perhaps, though, this is also one of the driving forces in my life: A yearning to improve myself so I can do things better "next time", assuming fate would give me another opportunity to do so.

In setting out to write this book, I did not intend for it to be some sort of sermon, or indeed a text book explaining my specialist field of nuclear magnetic resonance. It's simply a book about life, *my* life, about my search for the truth and my search for the true me. A lot of it may well be of interest just to me. Other parts – and this is my hope – will inspire others on their path to self-discovery. It's about the question of what we are all looking for, what are we trying to achieve, what do we want in life? Love? Self-respect? The respect of others? Wisdom even? Is it about the meaning of life or even some sort of immortality? Or are we simply being ruined by a disastrous craving for wealth, status, and fame?

I'd like to use the example of a parable, which originates in India. I have always admired the timeless nature of ancient Indian storytelling. There are thousands of stories that help pass on the eternal truths of life from one generation to the next. One parable that particular resonated with me was the "the three forgotten words":

"In a small village, not far from Adilabad in northern Andhra Pradesh, near to the geographical center of India, lived a farmer struggling for his meager living. Aged just 13, he was married to a girl from the neighboring village, at the time only eleven, who later bore him five children. Now all of his children had left home more than ten years ago, and unfortunately his wife had died three years previously. Our farmer was left alone with his cow and the two goats on a small piece of dry land, hardly able to survive.

Since his youth, he had constantly been asking questions about the world, about eternity, about the deeper meaning of life. His neighbors called him "the philosopher". He would frequently visit the famous Saraswati temple on the outskirts of Adilabad, asking Saraswati for enlightenment. He knew that the deity of wisdom would not respond to the naive questions of a simple farmer but, nevertheless, he hoped that her presence would help him to clarify the thoughts in his mind.

But the more he searched, the less he seemed to understand, and he became hopelessly desperate. On one dark night, he left his home; he untied his cow and goats, and let them experience freedom. He packed a small bundle of cloth and took the dusty road to the north. He walked for many days, asking people for directions to Mambhalam. Nobody had ever heard of such a village, but he was sure that there he would find the answers to all his burning questions. He walked for many weeks. He did not eat much and lived from berries that he found along the dusty road and in the scarce forests. He lost weight, and he reduced the size of his bundle to relieve his aching back. He walked for many months, and his steps became shorter and shorter. All too often, he had to sit down to regain his strength, but his will stayed alive, as he wanted to reach his goal. But more and more he worried that he would pass away before reaching Mambhalam.

One night, he was so desperate that he decided not to continue his journey any further, rather die here than walk and search in vain, being disappointed over and over again. He heard strange voices in the dark night and was frightened, feeling left alone. But tiredness overcame him and he fell asleep. He dreamed that he encountered an extremely old wise man, meditating naked under a huge tree. He sat down near to him and waited for a long time, for many days, and finally the old man opened his mouth, looked at his visitor with his twinkling eyes, and said just three words with a thundering voice. Then, suddenly he disappeared. The farmer woke up convinced that he had achieved the enlightenment he had been seeking for so long and felt immensely happy. But, unfortunately, he could no longer remember the three words. He thought as hard as he could, wracking his brain, and beating his head, but the three words did not come to him. He was close to utter despair.

The next day, he painfully started to realize that Mambhalam was actually located within himself, that he need not travel any further, but that he had to explore his own mind. He realized that searching in itself was indeed the true goal and purpose of life. He realized the connectivity of all causes and all phenomena. He realized that the entire world is united, as had been written in the holy scripts more than two thousand years ago. He learned to tame his unrest without suppressing it, turning it into a beneficial, fruitful direction. He found his happiness in the process of searching.

Later, many people came to visit him, asking for advice. He became a renowned teacher. He often told his students about the "dream of the three words". He encouraged them to find for themselves their own personal three words that would provide them a direction in life. He himself died an old man, and he is still remembered as the 'master of the three hidden secrets'. At the site where he died, the local people planted three trees. Since then, these trees have grown to become part of a sizable forest, and nobody remembers which ones were the original trees. The name of the wise man also has been forgotten. But inquisitive minds will continue his tradition of inquiry forever, not only in Andhra Pradesh, but all over the world."

I have been searching for my own "three words" of redemption for many, many years now. I have thought about hundreds of possible candidates, but I am yet to find any particular words I would consider worth committing to paper. I am still looking – and have long since realized that nobody will plant my three immortal trees for me.

How nuclear magnetic resonance came to benefit society

Diary entry, 1.1.96 (continued): What have I achieved recently? At work?

The question of how my research is relevant to society is one that actually occupied me my entire life. That is one of the reasons why following my dissertation, I left ETH and moved into industry. I wanted to do something that was actually of some tangible use. The ivory tower in which I spent much of my life was a terrible place to be. I had constant doubts as to the usefulness of my research, perhaps because the field of nuclear magnetic resonance is so theoretical that even when I was making my various discoveries, we had little or no idea how they could actually be applied in real life. I would never have expected, for example, that it would lead to the development of a process that would not only revolutionize science, but also healthcare for people around the world.

The history of nuclear magnetic resonance begins in the somewhat esoteric field of elementary particle physics. Certain nuclei are magnetic, and if they are exposed to radiowaves, they respond by sending back a weak signal of a certain frequency: What we call the resonance frequency. You'll hear back from them, you could say. At first, this discovery appeared to have little benefit for society as a whole – it was more something that captured the interest of just those physicists interested in the structure of atoms.

It was by pure coincidence that in 1950 scientists realized that the chemical environment has a shielding effect on the magnetic nuclei and that this is reflected in the NMR frequencies measured. It is as if the nuclei were leaving behind their "fingerprints"; you could also say that the melodies emanating from them revealed very specific information about their environment. In this way, the method became an essential tool for performing chemical analysis, both in academia and in the chemicals industry.

Experimental NMR has since become a real high-tech area. It requires extremely strong and stable magnetic fields, as the NMR signals emitted are unbelievably weak and state-of-the-art high-frequency electronic components are required to detect them. In addition to the hardware, extremely complex computer programs are required to analyze the highly-informative data generated by the process. NMR placed real demands on technological development.

This technological revolution was driven by the mathematical analysis of the measurements carried out on computers using the Fourier transform – which I together with Wes Anderson introduced at the start of the 1960s. This mathematical analysis process, which can also be found in the fields of optics and acoustics, converts a time-dependent signal into a frequency signal. This makes it possible to untangle complicated measurement results into simple frequency spectrums; it was possible to improve the sensitivity of the method by a factor of ten or even a hundred. By implication, this of course also makes it possible to perform ever more difficult experiments. This improvement in sensitivity was groundbreaking in terms of the method's application for complex biomolecules and use in the field of medicine.

Back at ETH Zurich, I worked together with my PhD students and postdocs as well as a number of other researchers to develop two- and three-dimensional NMR spectrums instead of the previous one-dimensional spectrums; here it is important to think of these dimensions not in terms of spatial dimensions, but in terms of time. By choosing the right sequences containing two or three precisely timed consecutive pulses, it is possible to excite the magnetic nuclei in such a way that they produce not just a simple "note", but rather a complex, polyphonic melody, which provides far more information about them. All we had to do was learn how to listen closely to these melodies. The breakthrough came from an idea developed on paper by Belgian physicist Jean Jeener, in combination with "our" untangling of ever more complex signal sequences using the Fourier transform.

It was later discovered that this expansion of NMR spectroscopy had a real impact on molecular biology, because it can be used to determine the three-dimensional (spatial) structure of biological macromolecules in the same state as they occur in nature. The information about molecular structures was essential when it came to examining the functioning of and interactions between biological substances, such as proteins as well as nucleic acids, which carry genetic material.

We later applied this expertise in the field of tomographic imaging, building on the idea put forward by scientists such as Paul Lauterbur, who laid the foundations for the use of NMR in medicine back in the early 1970s. As was the case with chemical analysis, we were able to design intelligent processes for exciting the nuclei and subsequently analyzing the results, all of which resulted in a marked improvement in terms of the speed and quality of the images. The idea was not yet ready for practical application and still needed to be developed and refined by other scientists. The first image of an entire head was not produced until 1978, while the first image of a whole body came in 1980 at the University of Aberdeen.

At first, scientists and experts still referred to the method as NMR imaging, but before the technology made it to the market, the name used in medical applications was changed to the current term "magnetic resonance imaging (MRI)". This was because at the time, anything to do with atomic nuclei was viewed with suspicion, despite NMR being completely safe. There are two major advantages to this technology: Firstly, it does not require the use of X-rays, which can indeed be harmful, so it means you can have as many MRI scans as necessary. Secondly, the quality of the images produced using MRI is far superior to X-rays. As the method involves measuring hydrogen nuclei, it is possible to see the body's soft tissue, that is to say the organs, blood vessels, muscles and ligaments, and not just the bones as is the case with X-rays. This makes it far easier to produce an exact diagnosis for diseases such as brain tumors, for example. You therefore can use MRI technology to identify almost all diseases and illnesses that change the body in some way. The method is unquestionably vital when it comes to diagnosing cancer, in particular.

Nowadays, MRI imaging is the most effective and universal diagnostic tool used by hospitals worldwide. In Swiss hospitals alone, there are almost 200 MRI scanners currently in use, each of which carry out more than 3,000 scans a year.

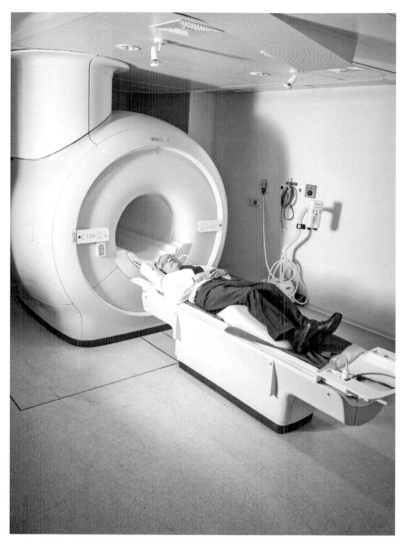

Richard Ernst in an MRI machine at Zurich University Hospital, photo taken on March 7, 2014. Ernst's inventions laid the foundations for today's widely used diagnostic method.

At the start of the 1990s, scientists such as Seiji Ogawa optimized the method even further. He developed the functional MRI (fMRI) process, which provides detailed and dynamic information about brain functions. It allows us to see the brain at work, so to speak.

We can now use the process to precisely locate most of the functions performed by the brain. This is of great use to psychologists, for example, who are able to closely examine human reactions and the way the various senses interact. Diagnostic markers have also been developed for numerous brain diseases. We can also expect to see another significant leap forward in the near future, which will further improve our understanding of the most complex and fascinating organ in the human body – the brain. MRI imaging is also playing an increasing role in the field of clinical psychology, making it possible to more closely research conditions such as autism, ADHS and other brain diseases.

The three pillars of responsible academia

Diary entry, 1.1.96 (continued): What have I achieved? In public life? I preach relevance – yet I've achieved nothing.

Without doubt, winning a Nobel Prize means that when you speak, others listen. But this didn't make my life easier – quite the opposite in fact. As I took this responsibility seriously, my life became ever more hectic. Barely a week passed when I was not somewhere in the world delivering a lecture, taking part in a conference, or meeting with various important people. I had to give presentations, was in demand as a consultant, and had to sit on various committees. I felt a bit like Albert Einstein – in 1921 he received the Nobel Prize for Physics – about whom the American author Burton Feldman told the following anecdote in his book "The Nobel Prize": Einstein was once asked by reporters to provide a definition of the fourth dimension, explain the theory of relativity in one sentence, give his views on prohibition, comment on the current political situation, explain what he thought about religion, and talk about his passion for the violin, all within a space of 15 minutes.

The thing that annoyed me the most, however, was the huge number of meetings I had to attend, all of which seemed to focus on petty power struggles and people's large egos. I have never held back from giving my opinion. Equally, I was not afraid of alienating people if I felt I was in the right. In Israel, for example, I criticized the government's occupation policy, while at a lecture in St. Petersburg, I did not mince my words when it came to discussing the superpowers' egotistical policies. On more than one occasion, my comments would result in an awkward silence, and once some of the audience made a point of leaving the room while I was still speaking.

Richard Ernst gives a speech on the occasion of the awarding of the prestigious Israeli Wolf Prize. He received it in 1991, shortly before the Nobel Prize.

I can't simply close my eyes to what is happening around the world. It evokes a sense of both sadness and anger within me. On the one hand, I see the beauty of nature, the kindness of many people, and the depth of human relationships that makes our lives so worth living. On the other hand, I am outraged by the destruction of the environment and the selfishness of man. Many people, particularly in western countries, live in a world of almost unparalleled abundance, while in other parts of the world, I see scenes of heartbreaking adversity and extreme poverty. As humans, do we not have a responsibility to rid our world of this inequality?

The reason for this lies in our excessive consumer society. Our daily lives are becoming more hectic from day to day. There are rivals lurking around every corner. To be successful, everyone has to keep one step ahead of the competition. We scientists are forced to come up with ever more inventions, publish our results more quickly in order to be viewed a success. Consumers are encouraged to buy and consume more and more things, all in the name of maintaining industrial productivity. The word "consumer" in itself is an ugly term, but is exactly right to describe the role that people nowadays fulfill. It is almost as if our purpose in life is to act as some sort of hungry black hole, sucking goods off the supermarket shelves as soon as they are put there. This then all results in mountains of waste and when we go to bed at night, we find ourselves wondering what the actual point is of all this hectic activity. We are unlikely to come up with a reassuring answer to this. But nevertheless, we wake up the next morning and climb back on to the eternal hamster wheel that is our consumer society.

Nowadays, people lack the humanity that could give their lives a deeper meaning. Their only remaining motivation – the only "megatrend" we see in society today – is to earn money. Indeed all forms of success are ultimately measured in terms of money. Like never before in the history of mankind, the dance around the golden calf, with all its excesses, has become a reality.

The consequence of this is that the environmental situation is becoming ever more hopeless all across the world. The overexploitation of nature and impending climate change worry me a great deal. Even today, there are obstinate scientists among us who maintain that global warming is an invention based on supposedly erroneous data. And then there are misguided high-ranking politicians who persist in putting forward this view. If you consider the facts, there is no other option but to conclude that the world is facing great danger.

We have to stop overexploiting our finite natural energy sources, such as oil and uranium, and instead focus on renewable sources, such as solar, wind, and geothermal power. And if these measures are not enough to feed our global energy consumption, then we must drastically cut back on the energy we do use. But the problem goes much deeper than this: I believe that with our current lifestyles, we are not only exhausting our natural resources, but we are acting against our conscience and human dignity. We are not only losing sight of the basis of our existence, but we are also losing the moral right to populate our increasingly depleted planet. If we carry on like this, we will soon be confronted with a global catastrophe that has the potential to wipe out humanity forever.

In its current state, the world is like a motorbike driving towards an abyss, on which we are all passengers. The rear wheel propels it forwards, driven by powerful industries whose aim is to maximize profits. Our politicians are sat on the seat, and their only job seems to be keeping consumer sentiment high so that they don't jump off the bike at full speed. They don't care in which direction the motorbike is traveling, even if it is heading for an abyss. The main thing is that the passengers spend as much of their money as possible on the goods produced by the factories and so boost their profits before the bike gets to its destination. In this situation, it is clear that somebody responsible has to take charge of the handlebars to make sure that the passengers are safe and turn the bike away from the abyss. Who would be better suited to take on this role than the academic community?

Who, if not we professors, with our privileged lives, should do this? Most people simply do not have the opportunity to do so or, if they raise a criticism, are ignored. I therefore believe that scientists, indeed the entire academic community, have a real responsibility to try and impact the world in a positive way. This responsibility is based on three pillars:

Education is our capital; it must be wide-ranging and not merely produce one-track specialists

Promoting education is by far the most important measure in a modern knowledge-based society. It allows people to become autonomous, self-realized beings and benefits all of humanity, as it has the potential to balance out inequalities, both within a single country as well as within a community of nations, for example between rich Northern Hemisphere countries and developing nations. It begins from a very early age. "Seeking knowledge at a young age is like engraving on a stone", wrote Islamic scholar Hasan al-Basri (642–728). The more developed and interconnected a society is, the more its peaceful existence and sustainable development depend on its investment in education. This also levels out inequalities, whether within a single social group or between different countries. That is why one of the most important responsibilities of a nation, and indeed the international community, is to provide its citizens with a proper education. I would even go so far as to say that it is *the* most important responsibility of a nation.

Education is much more than just teaching and learning things. Schools and particularly universities often focus almost exclusively on "teaching", that is to say passing on skills that are only of use in a specific job. Languages, science and mathematics, as well as manual skills and musical abilities, are only taught with the aim of turning students into able professionals – well-oiled cogs in society's complex machine.

I understand education to be something much more wide-ranging. In contrast to "teaching", a rounded "education" also shapes your personality and character. I am talking about a humanistic education based on explanation, much like the type of education called for by the famed German natural scientist Alexander von Humboldt. This does not just focus on learning things by rote from text books, but primarily on instilling certain values in students. It would not be limited to western cultures, as these values are in essence the same across all religions and philosophies, even if they are expressed in different ways. It is only by providing such an education that students will develop critical ways of thinking which will allow them to correctly assess the situation we find ourselves in.

Richard Ernst's NMR family, 1998. After his retirement, Richard invited all current and former employees to a meeting on Monte Verità lasting several days. Besides the social the researchers had many professional and interdisciplinary discussions.

Unfortunately many universities today focus almost exclusively on "teaching" instead of "education", especially when it comes to the exams that must be taken before obtaining a degree. They offer nothing more than a narrow, specialist education. How can we produce tomorrow's decision-makers, who will decide the fate of our planet, with this sort of education? The jobs that await these graduates require a far more wide-ranging, rounded education, one which takes the time to address ethical issues and social responsibility.

At a global level, the education of women is perhaps the most important issue. Anything that can be done to improve women's future prospects will benefit every single person on the planet. Women are the backbones of families and therefore society as a whole. If they are given the recognition they deserve, more resources and more freedom, the situation in their country is bound to improve. I think it is also the responsibility of the academic community to reconsider the roles of the various family members, if for no other reason than to achieve fairness.

This leads me to the next pillar of academic responsibility – the universities and their responsibilities as institutions to drive societal well-being.

Universities have a great responsibility

Our public universities can only have one mission: To think ahead about the future of the global community and to steer it in a positive direction. Politicians and economists are far too caught up in their short-term feedback cycles, whether this is about being reelected or generating sufficient return on investments. My trust in both politicians and economists has deteriorated significantly in recent years. Universities, by contrast, can and should enable their students to embrace a visionary spirit, to set themselves goals that go beyond day-to-day business and instead benefit the global community.

India's icon Mahatma Gandhi once came up with an unforgettable quote, one which has made a lasting impression on me: "You must be the change you wish to see in the world." These are brief, powerful words, but are not easily heeded. It is a lot easier to give other people "good" advice rather than consider your own attitude and way of thinking. But what does this all mean specifically for our universities?

Our academic institutions should (once again) become cultural centers that reach out into all of society – but I do not mean in an elitist way. We have to break out of these ivory towers and take more social responsibility! To do so, we must first of all break down the barriers that exist within the universities. We all know that we will need more than just technology and science to solve the major global problems. It is often the case that there are no technical hurdles to resolving a certain problem, but there are instead political, ethical or cultural barriers. To understand all of these, we need to draw on the humanities and social sciences. So let us break down the walls between the natural sciences, humanities and social sciences at our universities. Perhaps this will also open up the path to true wisdom.

At universities, we have to once again learn how to dream, how to create an ideal world, and how to implement our visions. We have a responsibility to discuss all the relevant issues in a manner that is as open and critical as possible. The influence exerted by politics and economics must be kept to a minimum. It is worth reminding ourselves that we are paid by the public to be open and critical in the representations we make, and it is thanks to our unique position that we are able to do so, even if blinkered politicians and unscrupulous lobbyists representing financial interests sometimes try to stop us from doing so. In a speech he gave in 1997 to celebrate the 200th birthday of writer Heinrich Heine, the then President of Germany Roman Herzog spoke about the inner freedom that intellectuals possess. "Without critical opposition," said Herzog, "without the commitment of those who express uncomfortable opinions, society would crumble." Society needs there to be thorns in the side of wealthy, we need critical thinkers not afraid to offer differing opinions, we need bold, rebellious minds and those who ask brave questions free of any conflict of interest. And it falls to us scientists, teachers and students at universities to take on this role!

Of course there is also a risk that these universities will degenerate into ivory towers. Academics can seem withdrawn or aloof, especially when scientific contact is limited to a narrow circle of other specialists. Evidence of this can frequently be found in the language used in scientific journals, for example. It is very much the case in the humanities that opaque specialist terms and unnecessarily complex lines of thought are used to explain what are often simple ideas. That is why it is so important that scientists follow up phases of teaching and research at universities with practical work, which may take place outside of the university. It is an absolute must for universities to work together with both public institutions and industry so that they can be more relevant in society.

Industry itself is not per se evil or exploitative. It serves an important purpose and is often the only social force able to convert scientific findings into actual products that benefit society as a whole. Lecturers and professors should therefore spend at least a year or so in a practical environment outside of academia. In turn, engineers and scientists from the industrial sector should regularly return to university to take on board new knowledge for themselves, on the one hand, and also to support students and academics with their practical experience and examples, on the other.

Having a multi-disciplinary education is an absolute necessity for anyone intending to work at the cutting edge of science. Nevertheless, scientists must also have an in-depth, detailed knowledge, at the very least of their own specialist area; all-rounders without this rarely achieve very much. You could summarize the situation as follows: While focusing on a specific field is necessary for understanding the subject, expanding your perspectives is necessary if you are to identify how the various fields interconnect.

This exchange of thoughts, ideas and experience must of course go beyond the walls of universities – it has to span different countries and continents. While science has for many centuries now had an international outlook, it has yet to embrace the potential offered by developing countries. It is in these regions, in particular, that we can clearly see the problematic side-effects of our materialistic consumer society. Without doubt, these developing nations will play a key role in determining the future course of humanity. It is only by experiencing and understanding the problems these countries face that it will be possible to come up with realistic ideas about a sustainable global future. That is why we have to maintain an active, open dialog with scientists and academics from these regions. But we must be careful not to merely impose our western model onto the countries of the South without considering all the key aspects – what may work well for us might not have any relevance elsewhere and so cannot be used as an example. That is why we must work together with thinkers from developing nations to come up with workable concepts for our future world.

One last important responsibility of universities is to pass on knowledge to society. Lecturers and researchers must spread their rational scientific findings, and also their critical opinions, around the world. It is only by doing so that we can restrict the destructive influence of false prophets and blinkered interest groups. This is the only way that society can take the wind out of the sails of the dangerous fundamentalism we see today. There are many types of fundamentalism. It ranges from fundamentalist Christian groups, who completely reject the science behind evolution and exert a disturbing amount of influence, particularly in the US, through to

Orthodox Judaism, Hindu intolerance, and Islamic fundamentalists with their destructive philosophy of violence. I view all of these groups with horror, as they replace critical, rational thought with a belief in simplified, unproven, and rigid dogmas. In my opinion, any form of fundamentalism can be seen as a type of "frozen ignorance", and one of the main goals of education is to "melt" this unshifting ignorance and help awaken a deeper understanding within people. Only then will there no longer be a place for fundamentalism. The true goal of university education – and indeed education as a whole – is to develop critical minds, paired with knowledge, wisdom, and tolerance, but with no tolerance for intolerance!

The third – and perhaps most important – pillar is the scientists themselves who work at the universities and institutions that produce the basic scientific and engineering knowledge:

Scientists must carry out research that is relevant and they cannot hide away from the public

We scientists are not rare shrubs that our society merely cultivates for the sake of it. Or in the words of Afghan Sufi sage Nawab Jan-Fishan Khan, "the candle is not there to illuminate itself." All academics have a mission and obligation to human society; it is this that justifies the significant amount of public money invested in our universities. We have more power than we think to influence the course of future global development. If universities are to become leading cultural centers and incubators of critical, forward-looking thought, this will be down to their staff being able to anticipate future trends. Many of us are in a privileged position, are well paid, and can do what we want. But this also means that we have to take our responsibilities seriously.

Firstly, scientists should work on issues that – at least over the long term – have some social benefit. Many people may say at this point that scientific research can only be free if it does not focus on single, specific goals with the aim of automatically producing meaningful applications. However, I believe that to be a poor excuse used by scientists to dodge the requirement to carry out relevant research, at least when their work is financed using public funding.

All scientists have a responsibility to think about the long-term consequences of their work and to incorporate these considerations into their planning.

Nevertheless, this requirement to benefit society must be interpreted in a sufficiently generous and far-sighted way. Results are not only relevant if they have some commercial use – research findings can also be culturally beneficial. Innovative findings not only form the basis for new technological developments, but can also enrich the lives of an inquisitive, culturally minded society.

Secondly, the work of researchers and scientists should span two levels at the same time. The first level relates to our original mandate, that is to say we must delve deeply into the secrets of nature and acquire the knowledge we need to solve a great many problems. This area of our work is based around the rules of fundamental scientific research. However, we must also bear in mind a second level, one that has a higher social and global significance. Here we have to consider a different perspective and look to understand our research in the context of society as a whole. It is within this higher level that social aspects and thus ethical considerations are extremely relevant. Nowadays, this applies to a great many fields, from energy research and the social sciences, through to genetic engineering.

The German philosopher Hans Jonas managed to describe ethical principles in such a neutral way that they are easy to accept even for scientists. In his 1984 work "The Imperative of Responsibility: In Search of an Ethics for the Technological Age", he proposed that the "projection of future consequences" should be used as a compass in deciding on the ethical application of scientific results. From this, he derived a definition of a scientist's responsibility: "Act so that the effects of your action are compatible with the permanence of genuine human life on Earth." This is actually the ultimate definition of sustainability! In line with Jonas' words, we have a duty to ensure that future generations enjoy the same opportunities that we ourselves had. We cannot continue exploiting non-renewable resources. It is clear that our currently way of living runs completely contrary to this imperative; we are living at the expense of other nations and future generations.

Even at a scientific university, social and ethical discourse should most definitely be included in the curriculum. There is no need for any special courses or outside programs for this. I would even go so far as to say we don't need any professors specializing in ethics. We cannot simply delegate or outsource the teaching of ethics in this way. The only thing we need to do is to appoint professors – no matter whether in the humanities or sciences – who have a vision of global responsibility, who are happy to integrate social and ethical issues into their science lectures as well. Another option here would be interdisciplinary discussion groups or summer schools, where lecturers and students can come together to examine and discuss these issues.

Thirdly, it is imperative that scientists and researchers ultimately open up a dialog with the general public. All scientists should be in a position to explain in clear terms the relevance of their projects to outsiders. It is only by being able to do so that we will continue to earn the trust of the public and thus continue to be funded by them. I am constantly amazed how little the general public, the press, and politicians know about scientific objectives and research. And this is mainly the fault of scientists themselves: It is our job to "advertise" our work. Yet many of us shy away from direct contact with the general public, claiming it is too time-consuming, that there is little direct benefit for us personally. But opening up a rift between science and the public can prove disastrous over the longer term. It affects the degree to which society feels obliged to support and fund our research, which over time will impede us in being able to resolve issues that are key to our future survival. We require support from all layers of society. Scientists have to become more active in terms of their communication, both with the general public via presentations, discussions and the press, but also with politicians, in particular. Communication is an important factor when it comes to the future of research, and is even more important for the future of society.

My guiding principle for academic responsibility is actually simple. We have an obligation to educate responsible, innovative managers of the future, who are prepared to work for the good of society. We need to produce the social pioneers of tomorrow. In his controversial, but influential pentalogy of novels *Gargantua and Pantagruel*, the French Renaissance poet and physician François Rabelais wrote that "Science sans conscience n'est que le ruine de l'âme", or science without a conscience will only ruin one's soul. This was at the beginning of the 16th century, when science in Europe was only just beginning to emerge from its deep slumber of the Middle Ages and free itself from the invisible, yet powerful religious shackles under which God the Almighty was responsible for everything.

The multi-faceted scientist

Diary entry, 1.1.96 (continued): What I am feeling is an emptiness. A creative emptiness?

Of course we scientists are only humans, even though we often have to suppress our feeling when it comes to putting in the hard work in the lab to discover objective facts. Yet when we grasp the potential of a new idea, develop a concept, or look for an approach to solve a problem we face, our emotions are extremely important to us as a source of inspiration and intuition. I am convinced that scientists who have gained experience in different, contrasting fields are often able to approach their work with a more intuitive approach. It is unlikely that a one-track specialist will make an earth-shattering discovery.

That is why it was always important for me to have a number of different interests outside of my research. My main passion outside of science has always been the arts: Whether the Tibetan art I discussed in an earlier chapter, or my love of classical music. The arts open up worlds in which you can experience the whole spectrum of human values, from cold, hard science, through to absurd reveries and overwhelming emotions that cannot be put into words. Since my early childhood, this space in my life has been filled by music. In fact, it was actually music's ability to convey the deepest human emotions that helped me to break out of the isolation I felt from my peers growing up and then to withstand the cloistered life of a hard-working, often desperate scientist. It is fortunate that my wife Magdalena also shares my passion. We would both have trouble coping without classical music, whether listening to it or actually playing it ourselves. My wife often used to sing in a choir and play the violin, while I learned the cello as a boy. Later on, I composed a number of short pieces and quickly became an avid concert-goer.

I soon became interested in contemporary experimental music and was fascinated by Charles Ives and his unconventional way of expressing himself musically, by Elliott Carter, John Cage, and George Antheil. I am convinced that for many scientists who have hidden themselves down their dark mineshafts away from the sunlight, music can be a way for them to get back in touch with their inner selves. When you listen to a complex composition, you can almost hear the meeting of the abstract sophistication of science with strong emotions.

Around five years ago on 30 May 2015, I had the opportunity as part of the "Music for a guest" series of concerts at the Musikkollegium in Winterthur, to put together my own concert to be played by the venue's orchestra. When I was approached about this in 2014, I knew straightaway that I would propose a program to honor the works of Johann Sebastian Bach, a composer that cannot be overlooked by any true fan of classical music. He is almost an omnipresent, invisible guest whenever you begin to discuss classical music; his sheer depth and range makes him an incomparable master of his art. For this concert, I also chose a number of works by later composers, which were all somehow related to Bach and so paid tribute to him in this way.

"My" concert opened with Anton Webern's "Ricercar a 6", his "opus 1". Webern was a Vienna-born composer and proponent of musical expressionism at the turn of the twentieth century. In this piece, Webern took an excerpt from Bach's "Musical Offering" and set it in six parts for an orchestra. This was followed by Bach's third Brandenburg Concerto, the second movement of which is typically omitted – I always found myself captivated by the piece's lively, rhythmic momentum. I had spent many years thinking about why

Bach would wanted to leave out the second movement. Did he expect that later musicians would then insert their own improvisations, or did he intentionally leave a gap so as to emphasize the contrast between the two quicker movements? I decided to take up Bach's challenge and replaced the missing second movement with a piece by American composer Charles Ives, one of my favorite composers: The Unanswered Question, a piece influenced by his surprising musical experimentation. Like Bach's Brandenburg Concerto, it is in G major, meaning it was an excellent fit. This combination as played by the Winterthur Orchestra was probably a world first – at this point, it is often the case that another short piece by Bach is played. It goes without saying that to do so would have been far too conventional for me.

After the interval, the outstanding pianist Karl-Andreas Kolly joined the orchestra to interpret Igor Stravinsky's powerful baroque "Concerto for Piano and Wind Instruments". The composer himself had performed this work almost eighty years earlier at the same venue in Winterthur, with his son Soulima Stravinsky on piano. I had been fascinated by this piece from an early age and had asked for a recording of it to mark my confirmation in 1949. The concerto has a sweeping dynamic, which is constantly interrupted by rhythmic changes – I think of it like a cog with a missing tooth. I saved the real highlight for the end: One of Johannes Brahms' major works, the last movement of his Fourth Symphony. The movement is based on Bach's cantata "Nach Dir, Herr, verlanget mir!" (For Thee, O Lord, I long) and is a compositional masterpiece.

The concert was a huge success and I got a lot of positive feedback. I was overjoyed to have had the opportunity to express myself musically in this setting.

Overall my life has been an emotional rollercoaster, full of highs and lows. Perhaps this is the same for any vaguely sensitive person, but I think that I am somewhat of an extreme case in this regard. I often felt myself torn between one extreme and the other – whenever I reached a peak in my life, I was always worried that this would be followed by an abrupt fall into the abyss below.

Richard Ernst sought in the science of physical chemistry always the secret that holds nature together at the innermost core.

After all, life is one big circle. Like a small wave that comes from throwing a stone into a calm pond, it slowly spreads, before dying down, and then ultimately disappearing completely. It is impossible to see where its energy has gone, whether it will go on to form new waves, or whether it was unique in the way it ebbs and flows. I cannot remember how I arrived in this world and it is unlikely that I will notice when I am about to leave. My life feels increasingly like I am sleepwalking, often waking up and no longer knowing how or where I fell asleep. It is like I am slowly departing this world – it is neither painful nor worrying. My only wish is for eternal rest to be granted unto me.

Epilogue by Prof. Alexander Wokaun

Since the 1970s, nuclear magnetic resonance has developed at a rapid pace. Its unprecedented ascent has impacted not only the area of chemical analysis, but also the ways in which we are now able to examine the structure of large, biologically significant macromolecules. The technique is used to analyze solids, and thanks to its ability to produce images using magnetic resonance tomography – MRI – it moved into the medical mainstream, with hospitals and radiologists relying on it to assist them in their diagnoses. The prominent scientist Richard Ernst helped lay the foundations for all of this and made a huge contribution to its development. It is therefore a real pleasure to hold this autobiography in my hands.

I got to know Richard Ernst in person for the first time when I was a chemistry undergraduate at ETH Zurich, where – as he describes in his memoirs – I was privileged to attend lectures given by professors of physical chemistry Hans Heinrich Günthard and Hans Primas. I was deeply impressed by Richard's lectures about measurement technology and the fundamentals of nuclear magnetic resonance, his personality, and the lively atmosphere in his lectures, so much that in early 1974, I turned to him to discuss the subject of my forthcoming Master's thesis. Richard Ernst assigned me a topic in the area of stochastic resonance, a field that is also briefly mentioned in his biographical notes.

In autumn of the same year, I obtained the opportunity to join his research group as a PhD student. During my time there, I came to know him not just as an eminent scientist, but also as a supervisor, mentor, and fatherly advisor. With an unshakeable commitment and inexhaustible engagement for science, he set himself the highest of standards, making him a true role model for everyone in the group, but without ever asking too much from any of us. I remember how we would often meet in the afternoon, discussing a scientific problem of some sort, and then the next morning, he would arrive with a neatly handwritten solution, which he had been working on late into the night. Our weekly group seminars may serve as another example – if a PhD student was not able to participate and present as foreseen, Richard would jump in, prepare, and talk about the subject of current interest himself.

The years between 1974 and 1978 were extremely exciting – it was during this time that two-dimensional spectroscopy laid the foundations for this technology to be adopted in the fields of chemistry, biology, and medicine, a step that was of extreme significance looking back today. In the labs around me, my colleagues from the group would work on increasing the sensitivity of the MRI process, and I got to experience first hand the excitement of their expectations and breakthroughs in imaging, while my own dissertation focused on other aspects of multi-dimensional spectroscopy. This marked a key milestone for me personally as I was able to complete my doctoral thesis.

This was the second time that having Richard as a mentor really benefited me – before spending time in the US during my postdoc, he advised me to focus on another experimental method. This saw me get involved in the field of laser spectroscopy, which went on later to form the basis of my habilitation work, carried out in a different team at ETH Zurich from 1982 to 1986. I also stayed in touch with Richard during this time, mainly thanks to our collaboration on a book – as always, he was the lively, upstanding scientist I had come to know.

Vividly I remember my excitement when in 1991 – at the time I was working at Bayreuth University – I heard the news that Richard had been awarded the Nobel Prize. While this meant a great deal to the scientific community in general, I was especially delighted on a personal level. I have fond memories of a seminar week in the southern Swiss canton of Ticino, to which Richard had invited all his former PhD students and staff to join him in looking back at the various developments made over those decisive years. It was astonishing to see how many different career paths the former members of the group had followed – this in itself was a sign of how well Richard's approach to providing us with a profound, multi-disciplinary education had prepared us all to embark upon our professional lives. He himself also had interests outside his scientific career – classical music, for example, or his passion for Tibetan art in particular. It was in this area that he became a real expert in the mythology and symbolism displayed in the Tibetan thangkas.

Winning the Nobel Prize marked the beginning of a new era of Richard Ernst's life devoted to science. He saw this award, which was more than deserved given what he had achieved in his career, more as an obligation to use the prominence he had gained from winning it to advance an issue that had always been close to his heart. Whether in his scientific papers or his lectures, he went on to highlight tirelessly the responsibility of scientists within society and the importance of reflecting on the purpose and consequences of our own discoveries. He embarked upon a relentless travel program, taking him all over the world and pushing him to his physical limits, in order to put forward these ideas at international conferences and research institutions to which he had been invited. Within the Swiss university system, he campaigned for administrative work to be kept to a minimum so as to free up scientists' time to develop new, innovative ideas. At the same time, he took very seriously his own administrative role at the university and mandate as an advisor on numerous committees, while continuing his scientific work with the same level of care and devotion. After he became an emeritus professor, he was able to dedicate more time to his interest in the history of arts, spending many hours analyzing the pigments used in his beloved thangkas, working out the dates and locations they were painted, and also learning how to restore paintings that had been damaged over time to their former brilliance.

Richard Ernst's decision to now write an autobiography, taking us through the many stages of his fascinating life, is something that I consider a gift to the scientific community – a gift that has been waited for with expectation. I am sure that anyone reading these vivid memoirs will enjoy and be captivated by them just as I was.

Alexander Wokaun, January 2021

Acknowledgments

This book would not have been possible were it not for the support and involvement of a great many people in my life. Whether I have known them for decades or our paths briefly crossed, they have all made an impression on me in one way or another and have a place in my heart.

First of all, I would like to thank my wife Magdalena, who has shown great patience in always being there for me, and my children Anna, Katharina and Hans Martin, whom I will always love and care for.

Thanks also go to my sisters Verena and Lisabet, my mother Irma Ernst-Brunner, and also my father Robert Ernst; while our father-son relationship may have been difficult, without him I am certain I would not have become who I am today.

I would also like to thank Roswith Tauber and her family; they had a positive influence on my life during my youth and I am delighted that we still remain friends to this day.

A very special thanks must go to Alexander Wokaun, who readily and generously spent many an hour going through this autobiography, checking and correcting any scientific facts that I may on occasions have misremembered due to my age.

Thank you also to Irène Müller, my loyal secretary, who has been there for me even after I retired, right through to this day, helping me maintain some order over my unruly filing and putting up with my sometimes challenging moods – thank you for your patience!

I would especially like to thank ETH Zurich as an institution and the Laboratory of Physical Chemistry, with all its staff, for being my scientific home for so many years. Thanks also go to my lecturers and colleagues for always sharing their ideas and opinions with me – without their support, I would never have achieved what I did. So, in no specific order, a big thank you goes to: Hans Heinrich Günthard and Hans Primas and my colleagues during my dissertation; Weston Anderson, former head of the science department at Varian Associates in Palo Alto, as well as my former colleagues in the lab in the Stanford Industrial Park; Felix Bloch, who was generous in sharing his vast scientific knowledge with me; Jean Jeener, whose revolutionary ideas gave my work new impetus during what were difficult times; my ETH colleague and fellow researcher Kurt Wüthrich, with whom I enjoyed an extremely productive partnership spanning a decade; the group around Horst Kessler and of course the scientists in my group at ETH Zürich, without whom it would have been completely impossible to achieve what we did, here in alphabetical order: Walter Aue, Gabriele Aebli, Peter Bachmann, Marc Baldus, Enrico Bartholdi, Thomas Baumann, Martin Blackledge, Geoffrey Bodenhausen, Serge Boentges, Lukas Braunschweiler, Rafael Brüschweiler, Tobias Bremi, Jacques Briand, Peter Brunner, Bernhard Brutscher, Douglas Burum, Pablo Caravatti, Mark L. Chu, Herman Cho, Christopher Counsell, Matthias Ernst (no relation to the author), Alexandra Frei, Zhehong Gan, Claudius Gemperle, Federico Graf, Christian Griesinger, Ron Haberkorn, Sabine Hediger, Eric Hoffmann, Alfred Höhener, Yongren Huang, Gunnar Jeschke, Jiri Karhan, Herbert Kogler, Roland Kreis, Anil Kumar, Hongbiao Le, Tilo Levante, Malcolm Levitt, Stephan Lienin, Max Linder, Slobodan Macura, Zoltan Madi, Andrew A. Maudsley, Mark McCoy, Beat H. Meier, Beat U. Meier, Rolf Meyer, Luciano Müller, Norbert Müller, Marcel Müri, Kuniaki Nagayama, Annalisa Pastore, Jeffrey W. Peng, Susanne Pfenniger, Christian Radloff, Mark Rance, David Redwine, Michael Reinhold, Günther Rist, Pierre Robyr, Thierry Schaffhauser, Stefan Schäublin, Christof Scheurer, Martin Schick-Pauli, Jürgen M. Schmidt, Paul Schosseler, Christian Schönenberger, Thomas Schulte-Herbrüggen, Arthur Schweiger, Jochen Sebbach, Gustavo Sierra, Nikolai R. Srynnikov, Scott Smith, Ole W. Sørensen, Armin Stöckli, Suzana Straus, Dieter Suter, Albert M. Thomas, Marco Tomaselli, Marcel Utz, René Verel, Thomas Wacker, Rico Wiedenbruch, Michael Willer, Dieter Welti, Ping Xu, Shanmin Zhang. My thanks also go to the hard-working staff in the electronic and mechanical workshops at the Institute for Physical Chemistry, who always provided outstanding, energetic support for our ideas for NMR experiments, no matter how impossible they appeared at the time.

I would also like to thank the staff at Bruker-Spectrospin for the fruitful partnership we enjoyed over so many years, specifically Tony Keller and Günther Laukien, who ran the company with foresight and no shortage of visionary strategies.

A special thanks must also go to all my friends and acquaintances who have helped me on my way outside of my scientific career. This includes the management team and staff at the Musikkollegium Winterthur, an institution I am pleased to have been able to support over the years, specifically the sadly deceased patron of the arts Werner Reinhart, as well as Maja Ingold, a patron and National Councillor from Winterthur. My warmest thanks also go to my friends at the Tibet Institute Rikon, specifically its founder Jacques Kuhn, as well as Rudolf Högger and longstanding Head of the Institute, Philipp Hepp.

If there is anyone I have forgotten to mention at this point, please do forgive me – this should in no way detract from your contribution or how thankful I am to you, but is more a reflection of my ever dwindling memory. Thank you.

Chronology

1933
Richard Robert Ernst born on 14 August to Robert Ernst and Irma Ernst-Brunner in Winterthur.

1934
Sister Verena born on 9 October.

1937
Sister Lisabet born on 15 July.

1940–1946
Attends primary school at the Schulhaus Inneres Lind in Winterthur.

1944
Isidor M. Rabi receives the Nobel Prize for Physics for his discovery of the magnetic moment of nuclei.

1946–1952
Attends the Im Lee High School in Winterthur.

1951
Internship at Hovag in Ems (now Ems-Chemie).

1952
Richard Ernst passes his high-school leaving exam. Felix Bloch and Edward Purcell receive the Nobel Prize for Physics for their discovery of nuclear magnetic resonance.

1952–1956
Undergraduate degree in the chemistry department at ETH Zurich.

1955

Industry internship in the science department of the dye division of Ciba under Heinrich Zollinger; first publication in the scientific journal *Helvetica Chimica Acta.*

1955

Father Robert Ernst dies on 2 August in Bad Nauheim aged 63.

1956

Richard Ernst completes his degree as a certified ETH Chemical Engineer.

1956/57

Officer training.

1957–1961

Dissertation at the Laboratory of Physical Chemistry of ETH Zurich under Hans Primas and Hans Heinrich Günthard.

1962

Submission of dissertation "*I. Kernresonanz-Spektroskopie mit stochastischen Hochfrequenzfeldern. II. Zur Konstruktion eines optimalen Kernresonanz-Messkopfes.* Dissertation no. 3300, ETH Zurich"; awarded the silver medal by ETH Zurich and a cash prize of 1,000 francs. Richard Ernst becomes a doctor of technical sciences at ETH Zurich.

1962/63

Researcher at the Laboratory of Physical Chemistry, ETH Zurich.

1963

Marriage of Richard Ernst and Magdalena Kielholz at the Oberwinterthur church on 5 October.

1963–1968

Emigrates to the USA; Richard Ernst works as a researcher in the Instrument division at Varian Associates in Palo Alto, California.

1964

Richard Ernst and Wes Anderson discover the pulse Fourier transform for NMR. Birth of daughter Anna Magdalena on 26 September.

1967
Birth of daughter Katharina Elisabeth on 26 May.

1968
Return to Switzerland via Asia, including Japan, Cambodia, Thailand, Nepal and India; first encounter with Tibetan scrolls (thangkas).

1968–1970
Richard Ernst becomes an outside lecturer in physical chemistry and head of research group at ETH Zurich; further development of Fourier transform.

1969
Richard Ernst awarded the Ružička Prize at ETH Zurich.

1970
Nervous breakdown and recuperation in Ticino in March/April.

1970–1972
Assistant professorship at ETH Zurich.

1971
Jean Jeener presents his fundamental ideas for a 2D experiment at the Ampère summer school.

1972
Birth of son Hans-Martin Walter on 9 December.

1972–1976
Associate professorship at ETH Zurich.

1974
Richard Ernst develops 2D NMR spectroscopy with the Fourier transform.

1976–1998
Full professorship at ETH Zurich; Richard Ernst further develops 2D NMR spectroscopy, computer analysis and Fourier NMR tomography for medical applications; research and development of various other NMR methods.

1976
Start of partnership with Kurt Wüthrich and his group.

1978
Moves in to new detached house in Winterthur.

1983
Richard Ernst receives the Gold Medal Award from the Society of Magnetic Resonance in Medicine, San Francisco.

1985
Honorary doctoratc from ETH Lausanne.

1986
Awarded Marcel Benoist Prize by the Swiss Confederation; end of partnership with the Wüthrich group.

1988
R.B. Woodward guest professor for chemistry at Harvard University in Cambridge, Massachusetts.

1988–2000
Member of Marcel Benoist Foundation committee.

1989
Richard Ernst receives the John Gamble Kirkwood medal from Yale University in New Haven, Connecticut, and the New Haven section of the American Chemical Society.

1989–2007
Vice-chairman of Board of Directors of Bruker BioSpin AG, Fällanden.

1990
Richard Ernst receives the Ampère Prize at the 25th Ampère Congress in Stuttgart.

1990–1994
President of ETH Zurich Research Commission.

1991
Richard Ernst receives the Wolf Prize for Chemistry, Jerusalem; the Louisa Gross Horwitz Prize for Biochemistry and Biology from Columbia University, New York; and the Nobel Prize for Chemistry, Stockholm.

1994
Richard Ernst awarded honorary doctorate by the Faculty of Medicine at the University of Zurich.

1997–2002
Member of the Swiss Science and Technology Council.

1998
Emeritus professor.

1998–2006
Member of University Council at the Technical University of Munich.

2002
Kurt Wüthrich receives the Nobel Prize in Chemistry for his NMR research into biomolecules.

2003
Paul C. Lauterbur and Sir Peter Mansfield receive the Nobel Prize in Medicine for their work in developing MRI imaging.

2004–2018
Member of Foundation Board of Tibet Institute Rikon.

2006
Mother Irma Ernst-Brunner dies on 21 May aged 97.

2009
Richard Ernst awarded honorary doctorate by the Faculty of Humanities at the University of Bern. This is his 17th honorary doctorate.

2021
Richard Ernst dies on June 4 in his hometown, Winterthur, aged 87.

Glossary

Atom

Atoms are the basic building blocks that make up matter. Each atom belongs to a particular chemical element. There are 92 naturally occurring elements and at least 24 artificially produced elements, all of which are methodically listed in the periodic table. The atoms making up the various elements differ in their structure (see atomic model).

Atomic model

There are various models that show how an atom is structured. However, they are all based around the assumption that an atom comprises a nucleus, which is made up of a varying number of positively charged protons and neutral neutrons depending on the element, as well as negatively charged electrons, which rotate around the nucleus. The nucleus of the atom is what matters when it comes to nuclear magnetic resonance methods.

Molecule

When several atoms are held together by chemical bonds, we have a molecule. An ethanol (alcohol) molecule, for example, comprises two carbon, six hydrogen and one oxygen atom (see Fig. 1). The type of bonds between the atoms and their spatial arrangement in the molecule is referred to by chemists as the structure of the molecule. It determines the chemical environment of the individual atoms within the molecule, which, for example, is different for a hydrogen atom bonded to an oxygen atom than it is for one bonded to a carbon atom.

Molecule

A molecule consists of atoms of different elements connected by chemical bonds.

Example:
Model of the ethanol molecule

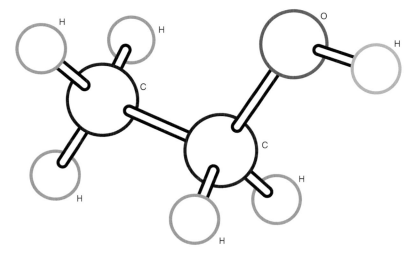

Ethanol consists of
1 oxygen atom (O, dark gray)
2 carbon atoms (C, black)
6 hydrogen atoms (H, light gray)

Structural formula: CH_3-CH_2-OH

Figure 1 Example of a simple molecule: Ethanol (alcohol).

Nuclear spin

Nuclear spin refers to the angular momentum generated by the rotation of the atomic nuclei; the nuclei therefore "spin" around themselves. All atoms have nuclear spin with the exception of those with an even number of protons *and* an even number of neutrons. The nuclei of hydrogen atoms (H), which consist of only one proton, are especially important when it comes to nuclear magnetic resonance.

Magnetic moment

If a nucleus has nuclear spin, the rotation of its electrical charge results in a magnetic moment. Such a nucleus then behaves like a small magnet with two opposing magnetic poles, which are influenced by magnetic fields and electromagnetic waves.

Frequency

Frequency refers to the number of occurrences per unit of time that a wave oscillates or a gyroscopic movement occurs, i.e. how quickly a recurring wave or gyroscopic movement passes through the same point. The base unit for measuring frequency is hertz, where 1 hertz corresponds to one oscillation or repetition per second.

Precession

If an external magnetic field is applied, the magnetic moment of the nucleus (comparable to a compass needle) aligns itself with respect to the axis of the magnetic field and rotates around the latter. This gyroscopic movement is known as precession motion. The precession speed of the magnetic nuclei has a certain frequency, the so-called Larmor frequency. It depends on the strength of the magnetic field applied, the properties of the nucleus and the chemical environment of the nucleus. For protons, the Larmor frequency for a magnetic field strength of 1.5 tesla is exactly 63.9 megahertz. By contrast, in the earth's magnetic field (approx. 30 microteslas at the equator and 60 microteslas at the poles) the Larmor frequency of a proton is between 1 and 2.5 kilohertz. By way of comparison, VHF radio signals are around 100 megahertz.

Radio waves

Radio waves are electromagnetic waves. They can be created by using an alternating current and have a certain wavelength and frequency (and therefore energy). With NMR methods, radio waves are directed onto the sample, either continually or in short pulses, and interact with the nuclear spin.

Excitation

When radio waves of the correct frequency are applied to a chemical substance, the magnetic nuclear spins absorb their energy, i.e. they become "excited".

Resonance frequency

During excitation, the rotating magnetic nuclear spins only absorb radio waves of a certain frequency; this is known as the resonance frequency. It corresponds exactly to the Larmor frequency (see precession) and as is the case with the Larmor frequency, it depends on the external magnetic field, the chemical environment and the properties of the nucleus measured. As per the rules of quantum chemistry, the nuclear spins reach a higher energy level when they absorb radiation at the resonance frequency.

NMR

Nuclear magnetic resonance (NMR) refers to the reaction of magnetic nuclear spins after being exposed to high-frequency radio waves. In chemical analysis, NMR is one of the best methods to examine the structure of molecules. The nuclei of hydrogen atoms (H) are especially important when it comes to nuclear magnetic resonance. However, NMR experiments can also be used to examine the nuclei of other suitable elements with nuclear spin.

NMR spectrometer

Device used to measure an NMR spectrum. The key component in an NMR spectrometer is the magnet together with the probe head as well as a radio wave transmitter and receiver, which is set to the resonance frequency range of the nucleus to be examined. There are also various auxiliary electronic devices up- and downstream from this, including the pre-amplifier, amplifier and modulators. A computer is used to control the experiment and then to analyze the results.

NMR experiment

The main magnet in an NMR spectrometer ensure that a strong and stable magnetic field is applied. The chemical substance to be analyzed is placed into the probe head in a special test tube between the pole shoes of the magnet. The strong magnetic field means that the nuclear spins of the hydrogen atoms in the molecule are aligned close to parallel (or antiparallel) to the magnetic field and precess around the axis of the magnetic field at their Larmor frequency (see Fig. 2). Then radio waves are sent to the sample via a transmitter. If the radio waves reach the resonance frequency of the nuclear spins, these are excited. After the radio waves are turned off, the excited nuclear spins return to their original state and re-release the energy they have absorbed, in the form of radio waves of the same frequency. Here, the nuclear spins induce an electrical signal in the receiver coil, which is then recorded. This makes it possible to measure an electromagnetic wave over a certain period of time; this is referred to as a time-dependent electromagnetic wave function. The wave function has a frequency, which corresponds exactly to the resonance frequency of the nucleus measured.

NMR spectrum

The NMR spectrum (also referred to as a "frequency spectrum", see Fig. 4) is a graphical representation of the various resonance frequencies of the nuclear spin measured in a molecule.

Chemical shift

In NMR experiments, the local magnetic field at the site of a hydrogen nucleus is influenced by the electromagnetic properties of the neighboring atoms in the molecule, in particular by those of the chemically bonded neighboring atoms. This chemical environment of the nuclear spin therefore changes the resonance frequency of the bare nucleus. In other words, from an NMR perspective there are various types of nuclear spins in a molecule depending on how these are bonded in the molecule. This results in the measured frequencies being shifted up or down by tiny amounts depending on the electrochemical properties of the neighboring atom. This effect is referred to as "chemical shift". It makes it possible to differentiate between and attribute the differently bonded hydrogen atoms within a molecule based on their resonance frequencies, and means that the structure of the entire molecule can be examined.

The NMR experiment

How the nuclear magnetic resonance method works

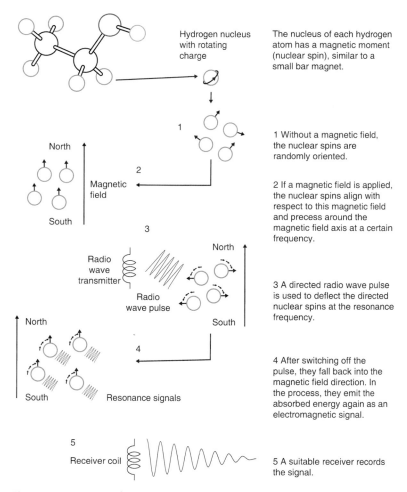

Hydrogen nucleus with rotating charge

The nucleus of each hydrogen atom has a magnetic moment (nuclear spin), similar to a small bar magnet.

North

South

Magnetic field

1 Without a magnetic field, the nuclear spins are randomly oriented.

2 If a magnetic field is applied, the nuclear spins align with respect to this magnetic field and precess around the magnetic field axis at a certain frequency.

Radio wave transmitter

Radio wave pulse

North

South

3 A directed radio wave pulse is used to deflect the directed nuclear spins at the resonance frequency.

North

South

Resonance signals

4 After switching off the pulse, they fall back into the magnetic field direction. In the process, they emit the absorbed energy again as an electromagnetic signal.

Receiver coil

5 A suitable receiver records the signal.

Figure 2 Structure of an NMR experiment.

Fourier transform

The Fourier transform (FT) is a mathematical operation, which determines the frequency of the time-dependent electromagnetic wave function. As the measured NMR signal is usually produced as a result of a variety of superimposed waves with different resonance frequencies, which can originate from the various different types of hydrogen nuclei within a molecule, these can be filtered out using the FT and represented in a frequency spectrum (see Fig. 3). However, the conversion process used in NMR is so complicated that it can only be carried out using powerful computers. The major benefit of the FT can be described by using an analogy: Imagine that you had to solve a complicated arithmetical problem in your head using roman numerals – a long-winded, tedious task. However, if you translate the roman numerals into the Arabic numerals we commonly use, the task becomes a whole lot easier. In this comparison, the process of translating of the roman numerals into the same Arabic numerals would take the role of the Fourier transform.

Pulse FT NMR method

A distinction is made between the continuous wave and the pulse FT NMR methods. With the continuous wave NMR method, the frequency of the radio waves sent to the sample is slowly changed so as to reach the resonance frequencies of the various nuclear spins one after the other. By contrast, with the pulse FT NMR method, a strong broadband radio wave pulse is sent to the sample, which covers the resonance frequencies of all the hydrogen nuclei in the molecule at the same time and allows these frequencies to be measured. This results in a superposition of signals, which can be decoded using the Fourier transform and converted into an NMR spectrum that can be easily interpreted. The advantage of the pulse FT NMR method is that it is up to 1,000 times faster than the previously used continuous wave method. Modern NMR equipment now only uses the pulse FT NMR method.

The Fourier transform

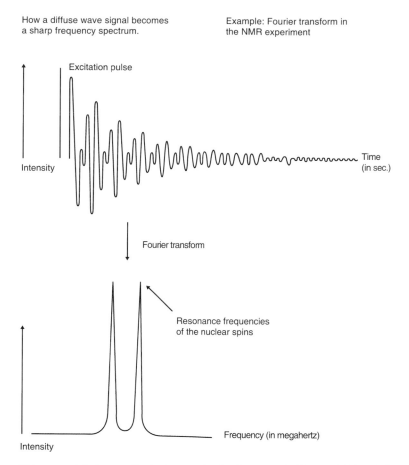

How a diffuse wave signal becomes a sharp frequency spectrum.

Example: Fourier transform in the NMR experiment

The measured electromagnetic signal, a wave function, is composed of the signal of all excited hydrogen atoms. By applying the Fourier transform, the individual resonance frequencies of the different nuclei can be filtered out and plotted along a frequency axis.

Figure 3   Example of a Fourier transform.

Noise

In principle, "noise" is the same in both the fields of acoustics and electronics: A mixture of waves or wavelets comprising electromagnetic or other waves with many different frequencies. We refer to "white noise" when it contains or is received at all the frequencies within a certain bandwidth. Because NMR experiments on a subatomic scale have to measure extremely low signal strengths, any background noise can disturb the signal.

Signal/Noise ratio

The ratio between the signal and the noise is a measure of how sensitive an NMR device is. The sensitivity of the measurement depends on how clean the sample is, the strength and stability of the magnetic field, and the way the signals are analyzed. Various technical processes are used to ensure optimum conditions. The test tubes can be rotated quickly, for example, to ensure the sample is mapped homogeneously. The magnetic field is stabilized and systemic fluctuations can be corrected retrospectively by using a computer program. Parts of the noise may be eliminated by electronic filters. All of this helps to optimize the signal/noise ratio. The experiment can also be repeated as often as required, with the results being calculated based on the average readings from all the experiments. The latter decisive improvement is achieved by applying the pulse FT method, which makes it possible to accumulate more experiments within a given total measurement time.

2D NMR

The two-dimensional NMR methods represents an enhancement of the procedure used to determine the structure of extremely large molecules comprising fifty or more atoms – natural proteins or DNA, for example. These molecules contain many different types of hydrogen atoms (with different chemical environments), which result in an extremely complex NMR spectrum with many overlapping resonance frequencies. A simple NMR measurement is therefore no longer sufficient. With the 2D NMR method, the hydrogen atoms in a molecule are excited with a radio wave pulse twice in succession. The response of the hydrogen atoms measured after these pulses differs according to their position in the molecule and their chemical environment, and depending on the time interval between the two pulses. The two dimensions refer to the time interval between the radio wave pulses, which is systematically varied as part of the experiment. The advantage of this double excitation is that the differences between the hydrogen atoms in terms of their chemical structure and position are being "amplified", which means that the various atoms and their positional and chemical relationships can be better identified.

The NMR spectrum

Where the clues to the structure of the molecule are hidden.

Example:
High-resolution NMR spectrum of ethanol (measurement of hydrogen nuclei)

Ethanol molecule
H: Hydrogen
C1: Carbon with three H atoms
C2: Carbon with two H atoms
O: Oxygen

Amplitude of the frequency signal: Indication of the number of atoms per molecule with the same chemical environment

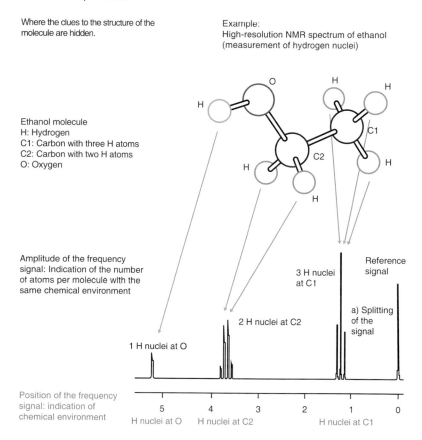

3 H nuclei at C1

Reference signal

2 H nuclei at C2

a) Splitting of the signal

1 H nuclei at O

Position of the frequency signal: indication of chemical environment

5   4   3   2   1   0
H nuclei at O   H nuclei at C2   H nuclei at C1

a) Splitting rule of frequency signals:
Indication of number of H nuclei bound to carbon atoms neighboring to the one considered.

Triplet signal: Each H-nucleus at C1 interacts with 2 H-nuclei each at the neighboring C2.
Quadruplet signal: Each H-nucleus at C2 interacts with 3 H-nuclei at the neighboring C1.
Singlet signal: H-nucleus at O has no interaction with other H-nuclei.

Figure 4   Model spectrum for ethanol and information that can be deduced from it.

Multidimensional NMR experiments

If three or more consecutive radio wave pulses (pulse sequences) are sent to a sample, the 2D NMR method can be expanded in various ways or into three or more dimensions.

MRI

MRI stands for magnetic resonance imaging (also referred to as magnetic resonance tomography, MRT). This medical imaging process is based on the same basic principle as the NMR method used in chemical analysis. An MRI device also aligns the nuclear spins of hydrogen atoms in a magnetic field and excites them using radio waves. In a body, it is mainly the hydrogen atoms found in water that are excited. However, an intricate technique is required to produce a three-dimensional image: In addition to a basic magnetic field, a linearly (gradually) increasing magnetic field is applied sequentially along each of the three physical axes, a "gradient". Because hydrogen atoms in a weaker magnetic field give off signals at a lower resonance frequency than atoms in a stronger field, the signal measured in a body can be precisely located. The MRI process measures one sectional plane after the other, and the results are then converted into a three-dimensional image. Various technical processes and optimizations, for which several Nobel Prizes have been awarded, mean the process is nowadays both quick and reliable.

Tissue differentiation

There are three parameters that determine the brightness of a tissue in an MRI image and therefore the contrast: 1. The density of the hydrogen atoms; 2. The time that passes between the nuclear spin receiving one pulse and it becoming excitable once more; and 3. The time it takes for the signal to fade away after the atom is excited. All three of these factors vary depending on the tissue, and it is this variance that opens up the diagnostic potential of the MRI method. When carrying out measurements, the parameters can be set such that the different properties are expressed to a greater or lesser degree and the organs can be clearly identified. This makes it possible, for example, to differentiate between white or gray brain matter, muscles, cartilage or even tumors and other tissues based on their characteristics without having to use contrast agents.

Functional MRI (fMRI)

The fMRI method represents a further development of the MRI method, which has allowed unimaginable progress to be made in the area of brain research, in particular. Japanese scientist Seiji Ogawa laid the foundations for this approach in the early 1990s. Unlike the MRI method, the fMRI method does not detect tissues or organs, but instead blood flows in the brain, given that increased brain activity increases oxygen demand in the relevant region. Hemoglobin is an iron-containing protein, which is loaded with oxygen as it passes through the lungs and then transports the oxygen in the blood. When a region of the brain is active, the ratio of oxygen-transporting hemoglobin to unloaded hemoglobin in the corresponding blood vessels is higher compared to inactive regions. It is possible to measure this difference as loaded hemoglobin has different magnetic properties to unloaded hemoglobin, and thus impacts the nuclear spin signals. This is known as the BOLD (blood oxygen level-dependent) effect.

Photo Credits

Page 70: Division of Work and Industry, National Museum of American History, Smithsonian Institution

Page 76: Wikimedia Commons

Page 102: ETH-Bibliothek Zürich, Bildarchiv

Page 114: 2D-H'NMR-Cosy-Spektrum, Prof. C. Griesinger

Page 144 (above): Felix Aeberli © StAAG/RBA13-RC02114-1_1

Page 144 (below): Felix Aeberli © StAAG/RBA13-RC02114-1_4

Page 146 (above): Felix Aeberli © StAAG/RBA13-RC02114-1_14

Pages 176 to 185: Sotheby's, Inc. © 2018

Page 194: Keystone-SDA/Rene Ruis

Pages 228, 232, 234, 236: © Simone Farner, Zürich

All other pictures: Privatarchiv Richard Ernst